Markov Chains: Models, Algorithms and Applications

Recent titles in the **INTERNATIONAL SERIES IN OPERATIONS RESEARCH & MANAGEMENT SCIENCE**
Frederick S. Hillier, Series Editor, *Stanford University*

* A list of the early publications in the series is at the end of the book *

Markov Chains: Models, Algorithms and Applications

Wai-Ki Ching   Michael K. Ng

Wai-Ki Ching
The University of Hong Kong
Hong Kong, P.R. China

Michael K. Ng
Hong Kong Baptist University
Hong Kong, P.R. China

ISBN-13: 978-1-4419-3986-2

e-ISBN-13: 978-0387-29337-0
e-ISBN-10: 0-387-29337-X

Printed on acid-free paper.

springeronline.com

To Anna, Cecilia, Mandy and our Parents

Contents

List of Figures

List of Tables

Preface

The aim of this book is to outline the recent development of Markov chain models for modeling queueing systems, Internet, re-manufacturing systems, inventory systems, DNA sequences, genetic networks and many other practical systems.

This book consists of eight chapters. In Chapter 1, we give a brief introduction to the classical theory on both discrete and continuous time Markov chains. The relationship between Markov chains of finite states and matrix theory will also be discussed. Some classical iterative methods for solving linear systems will also be introduced. We then give the basic theory and algorithms for standard hidden Markov model (HMM) and Markov decision process (MDP).

Chapter 2 discusses the applications of continuous time Markov chains to model queueing systems and discrete time Markov chain for computing the PageRank, the ranking of website in the Internet. Chapter 3 studies re-manufacturing systems. We present Markovian models for re-manufacturing, closed form solutions and fast numerical algorithms are presented for solving the systems. In Chapter 4, Hidden Markov models are applied to classify customers. We proposed a simple hidden Markov model with fast numerical algorithms for solving the model parameters. An application of the model to customer classification is discussed. Chapter 5 discusses Markov decision process for customer lifetime values. Customer Lifetime Values (CLV) is an important concept and quantity in marketing management. We present an approach based on Markov decision process to the calculation of CLV with practical data.

In Chapter 6, we discuss higher-order Markov chain models. We propose a class of higher-order Markov chain models with lower order of model parameters. Efficient numerical methods based on linear programming for solving the model parameters are presented. Applications to demand predictions, inventory control, data mining and DNA sequence analysis are discussed. In Chapter 7, multivariate Markov models are discussed. We present a class of multivariate Markov chain model with lower order of model parameters. Efficient numerical methods based on linear programming for solving the model parameters are presented. Applications to demand predictions and gene expression sequences are discussed. In Chapter 8, higher-order hidden Markov models are studies. We proposed a class of higher-order hidden Markov models with efficient algorithm for solving the model parameters.

This book is aimed at students, professionals, practitioners, and researchers in applied mathematics, scientific computing, and operational research, who are interested in the formulation and computation of queueing and manufacturing systems. Readers are expected to have some basic knowledge of probability theory Markov processes and matrix theory.

It is our pleasure to thank the following people and organizations. The research described herein is supported in part by RGC grants. We are indebted to many former and present colleagues who collaborated on the ideas described here. We would like to thank Eric S. Fung, Tuen-Wai Ng, Ka-Kuen Wong, Ken T. Siu, Wai-On Yuen, Shu-Qin Zhang and the anonymous reviewers for their helpful encouragement and comments; without them this book would not have been possible.

The authors would like to thank Operational Research Society, Oxford University Press, Palgrave, Taylor & Francis's and Wiley & Sons for the permissions of reproducing the materials in this book.

Hong Kong   *Wai-Ki CHING*
Hong Kong   *Michael K. NG*

1

Introduction

Markov chain is named after Prof. Andrei A. Markov (1856-1922) who first published his result in 1906. He was born on 14 June 1856 in Ryazan, Russia and died on 20 July 1922 in St. Petersburg, Russia. Markov enrolled at the University of St. Petersburg, where he earned a master's degree and a doctorate degree. He is a professor at St. Petersburg and also a member of the Russian Academy of Sciences. He retired in 1905, but continued his teaching at the university until his death. Markov is particularly remembered for his study of Markov chains. His research works on Markov chains launched the study of stochastic processes with a lot of applications. For more details about Markov and his works, we refer our reader to the following interesting website [220].

In this chapter, we first give a brief introduction to the classical theory on both discrete and continuous time Markov chains. We then present some relationships between Markov chains of finite states and matrix theory. Some classical iterative methods for solving linear systems will also be introduced. They are standard numerical methods for solving Markov chains. We will then give the theory and algorithms for standard hidden Markov model (HMM) and Markov decision process (MDP).

1.1 Markov Chains

This section gives a brief introduction to discrete time Markov chain. Interested readers can consult the books by Ross [180] and Häggström [103] for more details.

Markov chain concerns about a sequence of random variables, which correspond to the states of a certain system, in such a way that the state at one time epoch depends only on the one in the previous time epoch. We will discuss some basic properties of a Markov chain. Basic concepts and notations are explained throughout this chapter. Some important theorems in this area will also be presented.

Let us begin with a practical problem as a motivation. In a town there are two supermarkets only, namely Wellcome and Park'n. A marketing research indicated that a consumer of Wellcome may switch to Park'n in his/her next shopping with a probability of $\alpha(>0)$, while a consumer of Park'n may switch to Wellcome in his/her next shopping with a probability of $\beta(>0)$. The followings are two important and interesting questions. The first question is that what is the probability that a Wellcome's consumer will still be a Wellcome's consumer in his/her nth shopping? The second question is what will be the market share of the two supermarkets in the town in the long-run? An impoartant feature of this problem is that the future behavior of a consumer depends on his/her current situation. We will see later this marketing problem can be formulated by using a Markov chain model.

1.1.1 Examples of Markov Chains

We consider a stochastic process

$$\{X^{(n)}, n = 0, 1, 2, \ldots\}$$

that takes on a *finite* or *countable* set M.

Example 1.1. Let $X^{(n)}$ be the weather of the nth day which can be

$$M = \{sunny, windy, rainy, cloudy\}.$$

One may have the following realization:

$X^{(0)}$ =sunny, $X^{(1)}$ =windy, $X^{(2)}$ =rainy, $X^{(3)}$ =sunny, $X^{(4)}$ =cloudy, ...

Example 1.2. Let $X^{(n)}$ be the product sales on the nth day which can be

$$M = \{0, 1, 2, \ldots, \}.$$

One may have the following realization:

$$X^{(0)} = 4, \quad X^{(1)} = 5, \quad X^{(2)} = 2, \quad X^{(3)} = 0, \quad X^{(4)} = 5, \ldots.$$

Remark 1.3. For simplicity of discussion we assume M, the *state space* to be $\{0, 1, 2, \ldots\}$. An element in M is called a *state* of the process.

Definition 1.4. *Suppose there is a fixed probability P_{ij} independent of time such that*

$$P(X^{(n+1)} = i | X^{(n)} = j, X^{(n-1)} = i_{n-1}, \ldots, X^{(0)} = i_0) = P_{ij} \quad n \geq 0$$

where $i, j, i_0, i_1, \ldots, i_{n-1} \in M$. Then this is called a Markov chain process.

Remark 1.5. One can interpret the above probability as follows: the conditional distribution of any future state $X^{(n+1)}$ given the past states

$$X^{(0)}, X^{(2)}, \ldots, X^{(n-1)}$$

and present state $X^{(n)}$, is *independent* of the *past states* and *depends* on the *present state* only.

Remark 1.6. The probability P_{ij} represents the probability that the process will make a transition to state i given that currently the process is state j. Clearly one has

$$P_{ij} \geq 0, \quad \sum_{i=0}^{\infty} P_{ij} = 1, \quad j = 0, 1, \ldots.$$

For simplicity of discussion, in our context we adopt this convention which is different from the traditional one.

Definition 1.7. *The matrix containing P_{ij}, the transition probabilities*

$$P = \begin{pmatrix} P_{00} & P_{01} & \cdots \\ P_{10} & P_{11} & \cdots \\ \vdots & \vdots & \vdots \end{pmatrix}$$

is called the one-step transition probability matrix of the process.

Example 1.8. Consider the marketing problem again. Let $X^{(n)}$ be a 2-state process (taking values of $\{0, 1\}$) describing the behavior of a consumer. We have $X^{(n)} = 0$ if the consumer shops with Wellcome on the nth day and $X^{(n)} = 1$ if the consumer shops with Park'n on the nth day. Since the future state (which supermarket to shop in the next time) depends on the current state only, it is a Markov chain process. It is easy to check that the transition probabilities are

$$P_{00} = 1 - \alpha, \quad P_{10} = \alpha, \quad P_{11} = 1 - \beta \quad \text{and} \quad P_{01} = \beta.$$

Then the one-step transition matrix of this process is given by

$$P = \begin{pmatrix} 1 - \alpha & \beta \\ \alpha & 1 - \beta \end{pmatrix}.$$

Example 1.9. (Random Walk) Random walks have been studied by many physicists and mathematicians for a number of years. Since then, there have been a lot of extensions [180] and applications. Therefore it is obvious for discussing the idea of random walks here. Consider a person who performs a random walk on the real line with the counting numbers

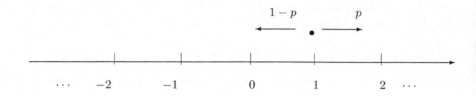

Fig. 1.1. The random walk.

$$\{\ldots, -2, -1, 0, 1, 2, \ldots\}$$

being the state space, see Fig. 1.1. Each time the person at state i can move one step forward (+1) or one step backward (-1) with probabilities p $(0 < p < 1)$ and $(1 - p)$ respectively. Therefore we have the transition probabilities

$$P_{ji} = \begin{cases} p & \text{if } j = i + 1 \\ 1 - p & \text{if } j = i - 1 \\ 0 & \text{otherwise.} \end{cases}$$

for $i = 0, \pm1, \pm2, \ldots$.

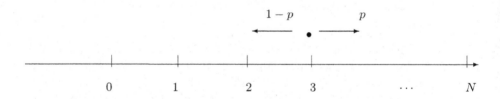

Fig. 1.2. The gambler's problem.

Example 1.10. (Gambler's Ruin) Consider a gambler gambling in a series of games, at each game, he either wins one dollar with probability p or loses one dollar with probability $(1 - p)$. The game ends if either he loses all his money or he attains a total amount of N dollars. Let the gambler's fortune be the state of the gambling process then the process is a Markov chain. Moreover, we have the transition probabilities

$$P_{ji} = \begin{cases} p & \text{if } j = i + 1 \\ 1 - p & \text{if } j = i - 1 \\ 0 & \text{otherwise.} \end{cases}$$

for $i = 1, 2, \ldots, N-1$ and $P_{00} = P_{NN} = 1$. Here state 0 and N are called the *absorbing states*. The process will stay at 0 or N forever if one of the states is reached.

1.1.2 The nth-Step Transition Matrix

In the previous section, we have defined the one-step transition probability matrix P for a Markov chain process. In this section, we are going to investigate the n-step transition probability $P_{ij}^{(n)}$ of a Markov chain process.

Definition 1.11. *Define $P_{ij}^{(n)}$ to be the probability that a process in state j will be in state i after n additional transitions. In particular $P_{ij}^{(1)} = P_{ij}$.*

Proposition 1.12. $P^{(n)} = P^n$ *where $P^{(n)}$ is the n-step transition probability matrix and P is the one-step transition matrix.*

Proof. We will prove the proposition by using mathematical induction. Clearly the proposition is true when $n = 1$. We then assume that the proposition is true for n. We note that

$$P^n = \underbrace{P \times P \times \ldots \times P}_{n \ times}.$$

Then

$$P_{ij}^{(n+1)} = \sum_{k \in M} P_{ki}^{(n)} P_{jk}^{(1)} = \sum_{k \in M} P_{ki}^n P_{jk} = [P^{n+1}]_{ij}.$$

By the principle of mathematical induction the proposition is true for all non-negative integer n.

Remark 1.13. It is easy to see that

$$P^{(m)} P^{(n)} = P^m P^n = P^{m+n} = P^{(m+n)}.$$

Example 1.14. We consider the marketing problem again. In the model we have

$$P = \begin{pmatrix} 1-\alpha & \beta \\ \alpha & 1-\beta \end{pmatrix}.$$

If $\alpha = 0.3$ and $\beta = 0.4$ then we have

$$P^{(4)} = P^4 = \begin{pmatrix} 0.7 & 0.4 \\ 0.3 & 0.6 \end{pmatrix}^4 = \begin{pmatrix} 0.5749 & 0.5668 \\ 0.4351 & 0.4332 \end{pmatrix}.$$

Recall that a consumer is in state 0 (1) if he/she is a consumer of Wellcome (Park'n). $P_{00}^{(4)} = 0.5749$ is the probability that a Wellcome's consumer will

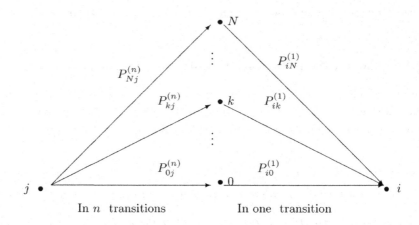

Fig. 1.3. The $(n + 1)$-step transition probability.

shop with Wellcome on his/her fourth shopping and $P_{10}^{(4)} = 0.4351$ is the probability that a Wellcome's consumer will shop with Park'n on his/her fourth shopping. $P_{01}^{(4)} = 0.5668$ is the probability that a consumer of Park'n will shop with Wellcome on his/her fourth shopping. $P_{11}^{(4)} = 0.4332$ is the probability that a consumer of Park'n will shop with Park'n on his/her fourth shopping.

Remark 1.15. Consider a Markov chain process having states in $\{0, 1, 2, \ldots\}$. Suppose that we are given at time $n = 0$ the probability that the process is in state i is $a_i, i = 0, 1, 2, \ldots$. One interesting question is the following. What is the probability that the process will be in state j after n transitions? In fact, the probability that given the process is in state i and it will be in state j after n transitions is $P_{ji}^{(n)} = [P^n]_{ji}$, where P_{ji} is the one-step transition probability from state i to state j of the process. Therefore the required probability is

$$\sum_{i=0}^{\infty} P(X^{(0)} = i) \times P_{ji}^{(n)} = \sum_{i=0}^{\infty} a_i \times [P^n]_{ji}.$$

Let

$$\mathbf{X}^{(n)} = (\tilde{X}_0^{(n)}, \tilde{X}_1^{(n)}, \ldots,)$$

be the probability distribution of the states in a Markov chain process at the nth transition. Here $\tilde{X}_i^{(n)}$ is the probability that the process is in state i after n transitions and

$$\sum_{i=0}^{\infty} \tilde{X}_i^{(n)} = 1.$$

It is easy to check that

$$\mathbf{X}^{(n+1)} = P\mathbf{X}^{(n)}$$

and

$$\mathbf{X}^{(n+1)} = P^{(n+1)}\mathbf{X}^{(0)}.$$

Example 1.16. Refer to the previous example. If at $n = 0$ a consumer belongs to Park'n, we may represent this information as

$$\mathbf{X}^{(0)} = (\tilde{X}_0^{(0)}, \tilde{X}_1^{(0)})^T = (0, 1)^T.$$

What happen on his/her fourth shopping?

$$\mathbf{X}^{(4)} = P^{(4)}\mathbf{X}^{(0)} = \begin{pmatrix} 0.7 & 0.4 \\ 0.3 & 0.6 \end{pmatrix}^4 (0, 1)^T = (0.5668, 0.4332)^T.$$

This means that with a probability 0.4332 he/she is still a consumer of Park'n and a probability 0.5668 he/she is a consumer of Wellcome on his/her fourth shopping.

1.1.3 Irreducible Markov Chain and Classifications of States

In the following, we define two definitions for the states of a Markov chain.

Definition 1.17. *In a Markov chain, state i is said to be reachable from state j if $P_{ij}^{(n)} > 0$ for some $n \geq 0$. This means that starting from state j, it is possible (with positive probability) to enter state i in finite number of transitions.*

Definition 1.18. *State i and state j are said to communicate if state i and state j are reachable from each other.*

Remark 1.19. The definition of communication defines an equivalent relation.
(i) state i communicates with state i in 0 step because

$$P_{ii}^{(0)} = P(X^{(0)} = i | X^{(0)} = i) = 1 > 0.$$

(ii)If state i communicates with state j, then state j communicates with state i.
(iii)If state i communicates with state j and state j communicates with state k then state i communicates with state k. Since $P_{ji}^{(m)}, P_{kj}^{(n)} > 0$ for some m and n, we have

$$P_{ki}^{(m+n)} = \sum_{h \in M} P_{hi}^{(m)} P_{kh}^{(n)} \geq P_{ji}^{(m)} P_{kj}^{(n)} > 0.$$

Therefore state k is reachable from state i. By inter-changing the roles of i and k, state i is reachable from state k. Hence i communicates with k. The proof is then completed.

Definition 1.20. *Two states that communicates are said to be in the same class. A Markov chain is said to be irreducible, if all states belong to the same class, i.e. they communicate with each other.*

Example 1.21. Consider the transition probability matrix

$$
\begin{matrix} 0 \\ 1 \\ 2 \end{matrix}
\begin{pmatrix} 0.0\ 0.5\ 0.5 \\ 0.5\ 0.0\ 0.5 \\ 0.5\ 0.5\ 0.0 \end{pmatrix}
$$

Example 1.22. Consider another transition probability matrix

$$
\begin{matrix} 0 \\ 1 \\ 2 \\ 3 \end{matrix}
\begin{pmatrix} 0.0\ 0.0\ 0.0\ 0.0 \\ 1.0\ 0.0\ 0.5\ 0.5 \\ 0.0\ 0.5\ 0.0\ 0.5 \\ 0.0\ 0.5\ 0.5\ 0.0 \end{pmatrix}.
$$

We note that from state 1, 2, 3, it is not possible to visit state 0, i.e

$$
P_{01}^{(n)} = P_{02}^{(n)} = P_{03}^{(n)} = 0.
$$

Therefore the Markov chain is not irreducible (or it is reducible).

Definition 1.23. *For any state i in a Markov chain, let f_i be the probability that starting in state i, the process will ever re-enter state i. State i is said to be recurrent if $f_i = 1$ and transient if $f_i < 1$.*

We have the following proposition for a recurrent state.

Proposition 1.24. *In a finite Markov chain, a state i is recurrent if and only if*

$$
\sum_{n=1}^{\infty} P_{ii}^{(n)} = \infty.
$$

By using Proposition (1.24) one can prove the following proposition.

Proposition 1.25. *In a finite Markov chain, if state i is recurrent (transient) and state i communicates with state j then state j is also recurrent (transient).*

1.1.4 An Analysis of the Random Walk

Recall the classical example of random walk, the analysis of the random walk can also be found in Ross [180]. A person performs a random walk on the real line of integers. Each time the person at state i can move one step forward (+1) or one step backward (-1) with probabilities p $(0 < p < 1)$ and $(1 - p)$ respectively. Since all the states are communicated, by Proposition 1.25, all states are either recurrent or they are all transient.

Let us consider state 0. To classify this state one can consider the following sum:

$$\sum_{m=1}^{\infty} P_{00}^{(m)}.$$

We note that

$$P_{00}^{(2n+1)} = 0$$

because in order to return to state 0, the number of forward movements should be equal to the number of backward movements and therefore the number of movements should be even and

$$P_{00}^{(2n)} = \binom{2n}{n} p^n (1-p)^n.$$

Hence we have

$$I = \sum_{m=1}^{\infty} P_{00}^{(m)} = \sum_{n=1}^{\infty} P_{00}^{(2n)} = \sum_{n=1}^{\infty} \binom{2n}{n} p^n (1-p)^n = \sum_{n=1}^{\infty} \frac{(2n)!}{n!n!} p^n (1-p)^n.$$

Recall that if I is finite then state 0 is transient otherwise it is recurrent. Then we can apply the Stirling's formula to get a conclusive result. The Stirling's formula states that if n is large then

$$n! \approx n^{n+\frac{1}{2}} e^{-n} \sqrt{2\pi}.$$

Hence one can approximate

$$P_{00}^{(2n)} \approx \frac{(4p(1-p))^n}{\sqrt{\pi n}}.$$

There are two cases to consider. If $p = \frac{1}{2}$ then we have

$$P_{00}^{(2n)} \approx \frac{1}{\sqrt{\pi n}}.$$

If $p \neq \frac{1}{2}$ then we have

$$P_{00}^{(2n)} \approx \frac{a^n}{\sqrt{\pi n}}$$

where

$$0 < a = 4p(1-p) < 1.$$

Therefore when $p = \frac{1}{2}$, state 0 is recurrent as the sum is infinite, and when $p \neq \frac{1}{2}$, state 0 is transient as the sum is finite.

1.1.5 Simulation of Markov Chains with EXCEL

Consider a Markov chain process with three states $\{0, 1, 2\}$ with the transition probability matrix as follows:

$$P = \begin{matrix} 0 \\ 1 \\ 2 \end{matrix} \begin{pmatrix} 0.2 & 0.5 & 0.3 \\ 0.3 & 0.1 & 0.3 \\ 0.5 & 0.4 & 0.4 \end{pmatrix}.$$

Given that $X_0 = 0$, our objective here is to generate a sequence

$$\{X^{(n)}, n = 1, 2, \ldots\}$$

which follows a Markov chain process with the transition matrix P.

To generate $\{X^{(n)}\}$ there are three possible cases:

(i) Suppose $X^{(n)} = 0$, then we have

$$P(X^{(n+1)} = 0) = 0.2 \qquad P(X^{(n+1)} = 1) = 0.3 \qquad P(X^{(n+1)} = 2) = 0.5;$$

(ii) Suppose $X^{(n)} = 1$, then we have

$$P(X^{(n+1)} = 0) = 0.5 \qquad P(X^{(n+1)} = 1) = 0.1 \qquad P(X^{(n+1)} = 2) = 0.4;$$

(iii) Suppose $X^{(n)} = 2$, then we have

$$P(X^{(n+1)} = 0) = 0.3 \qquad P(X^{(n+1)} = 1) = 0.3 \qquad P(X^{(n+1)} = 2) = 0.4.$$

Suppose we can generate a random variable U which is uniformly distributed over $[0, 1]$. Then one can generate the distribution in Case (i) when $X^{(n)} = 0$ easily as follows:

$$X^{(n+1)} = \begin{cases} 0 & \text{if} \quad U \in [0, 0.2), \\ 1 & \text{if} \quad U \in [0.2, 0.5), \\ 2 & \text{if} \quad U \in [0.5, 1]. \end{cases}$$

The distribution in Case (ii) when $X^{(n)} = 1$ can be generated as follows:

$$X^{(n+1)} = \begin{cases} 0 & \text{if} \quad U \in [0, 0.5), \\ 1 & \text{if} \quad U \in [0.5, 0.6), \\ 2 & \text{if} \quad U \in [0.6, 1]. \end{cases}$$

The distribution in Case (iii) when $X^{(n)} = 2$ can be generated as follows:

$$X^{(n+1)} = \begin{cases} 0 & \text{if} \quad U \in [0, 0.3), \\ 1 & \text{if} \quad U \in [0.3, 0.6), \\ 2 & \text{if} \quad U \in [0.6, 1]. \end{cases}$$

In EXCEL one can generate U, a random variable uniformly distributed over $[0, 1]$ by using "=rand()". By using simple logic statement in EXCEL, one can simulate a Markov chain easily. The followings are some useful logic statements in EXCEL used in the demonstration file.

(i) "B1" means column B and Row 1.
(ii) "=IF(B1=0,1,-1)" gives 1 if B1=0 otherwise it gives -1.
(iii) "=IF(A1 > B2,0,1)" gives 0 if A1 > B2 otherwise it gives 1.
(iv) "=IF(AND(A1=1,B2>2),1,0)" gives 1 if A1=1 and B2>2 otherwise it gives 0.
(v) "=max(1,2,-1) =2 " gives the maximum of the numbers.

A demonstration EXCEL file is available at [221] for reference. The program generates a Markov chain process

$$X^{(1)}, X^{(2)}, \ldots, X^{(30)}$$

whose transition probability is P and $X^{(0)} = 0$.

1.1.6 Building a Markov Chain Model

Given an observed data sequence $\{X^{(n)}\}$, one can find the transition frequency F_{jk} in the sequence by counting the number of transitions from state j to state k in one step. Then one can construct the one-step transition matrix for the sequence $\{X^{(n)}\}$ as follows:

$$F = \begin{pmatrix} F_{11} & \cdots\cdots & F_{1m} \\ F_{21} & \cdots\cdots & F_{2m} \\ \vdots & \vdots\ \vdots & \vdots \\ F_{m1} & \cdots\cdots & F_{mm} \end{pmatrix}. \tag{1.1}$$

From F, one can get the estimates for P_{jk} as follows:

$$P = \begin{pmatrix} P_{11} & \cdots\cdots & P_{1m} \\ P_{21} & \cdots\cdots & P_{2m} \\ \vdots & \vdots\ \vdots & \vdots \\ P_{m1} & \cdots\cdots & P_{mm} \end{pmatrix} \tag{1.2}$$

where

$$P_{jk} = \begin{cases} \dfrac{F_{jk}}{\sum\limits_{j=1}^{m} F_{jk}} & \text{if } \sum\limits_{j=1}^{m} F_{jk} > 0 \\[4mm] 0 \text{ if } \sum\limits_{j=1}^{m} F_{jk} = 0. \end{cases}$$

We consider a sequence $\{X^{(n)}\}$ of three states ($m = 3$) given by

"U" is a column of random numbers in (0,1). Column E (J) [O] gives the the next state given that the current state is 0 (1) [2]. Column P gives the simulated sequence X(t) given that X(0)=0.

U	0	1	2	X(t+1)|X(t)=0	U	0	1	2	X(t+1)|X(t)=1	U	0	1	2	X(t+1)|X(t)=2	X(t) 0
0.55	-1	-1	2	2	0.065	-1	1	-1	1	0.82	-1	1	-1	2	2
0.74	-1	-1	2	2	0.523	-1	-1	2	2	0.96	-1	-1	2	1	1
0.72	-1	-1	2	2	0.55	-1	-1	2	2	0.18	-1	-1	2	2	2
1	-1	-1	2	2	0.34	-1	-1	2	2	0.42	-1	-1	2	2	2
0.96	-1	-1	2	2	0.92	-1	-1	2	2	0.91	-1	-1	2	2	2
0.25	-1	1	-1	1	0.593	0	-1	-1	0	0.05	0	-1	-1	2	2
0.83	-1	-1	2	2	0.377	-1	-1	2	2	0.74	-1	-1	2	0	0
0.97	-1	-1	2	2	0.09	-1	-1	2	2	0.41	-1	-1	2	2	2
0.91	-1	-1	2	2	0.682	-1	-1	2	2	0.38	-1	-1	2	2	2
0.5	-1	-1	2	2	0.198	-1	1	-1	1	0.68	-1	1	-1	2	2
0.26	-1	1	-1	1	0.52	0	-1	-1	0	0.61	0	-1	-1	1	1
0.76	-1	-1	2	2	0.884	-1	-1	2	2	0.13	-1	-1	2	0	2
0.35	-1	1	-1	1	0.769	0	-1	-1	0	0.55	-1	1	-1	2	2
0.92	-1	-1	2	2	0.286	-1	-1	2	2	0.98	-1	-1	2	1	1
0.57	-1	-1	2	2	0.436	-1	1	-1	1	0.27	-1	1	-1	2	1
0.11	0	-1	-1	0	0.421	0	-1	-1	0	0.45	0	-1	-1	1	0
0.85	-1	-1	2	2	0.938	-1	-1	2	2	0.07	-1	-1	2	0	2
0.11	0	-1	-1	0	0.695	0	-1	-1	0	0.08	0	-1	-1	2	2
0.06	0	-1	-1	0	0.622	0	-1	-1	0	0.18	0	-1	-1	0	0
0.21	-1	1	-1	1	0.44	0	-1	-1	0	0.87	0	-1	-1	0	1
0.58	-1	-1	2	2	0.081	-1	1	-1	1	0.52	-1	1	-1	0	1
0.82	-1	-1	2	2	0.358	-1	-1	2	2	0.49	-1	-1	2	1	2
0.98	-1	-1	2	2	0.685	-1	-1	2	2	0.24	-1	-1	2	2	2
0.8	-1	-1	2	2	0.691	-1	-1	2	2	0.11	-1	-1	2	2	2
0.81	-1	-1	2	2	0.138	-1	-1	2	2	0.99	-1	-1	2	2	2
0.52	-1	-1	2	2	0.1	-1	1	-1	1	0.61	-1	1	-1	2	2
0.16	0	-1	-1	0	0.713	0	-1	-1	0	0.97	0	-1	-1	1	1
0.22	-1	1	-1	1	0.54	0	-1	-1	0	0.48	0	-1	-1	0	0
0.19	0	-1	-1	0	0.397	0	-1	-1	0	0.18	0	-1	-1	0	0
0.64	-1	-1	2	2	0.673	-1	-1	2	2	0.09	-1	-1	2	0	2

Fig. 1.4. Simulation of a Markov chain.

$$\{0,0,1,1,0,2,1,0,1,2,0,1,2,0,1,2,0,1,0,1\}. \tag{1.3}$$

We have the transition frequency matrix

$$F = \begin{pmatrix} 1 & 3 & 3 \\ 6 & 1 & 1 \\ 1 & 3 & 0 \end{pmatrix}. \tag{1.4}$$

Therefore one-step transition matrices can be estimated as follows:

$$P = \begin{pmatrix} 1/8 & 3/7 & 3/4 \\ 3/4 & 1/7 & 1/4 \\ 1/8 & 3/7 & 0 \end{pmatrix}. \tag{1.5}$$

A demonstration EXCEL file is available at [222] for reference.

X(t)	P00	P01	P02	P10	P11	P12	P20	P21	P22
0	1	0	0	0	0	0	0	0	0
0	0	1	0	0	0	0	0	0	0
1	0	0	0	0	1	0	0	0	0
1	0	0	0	1	0	0	0	0	0
0	0	0	1	0	0	0	0	0	0
2	0	0	0	0	0	0	0	1	0
1	0	0	0	1	0	0	0	0	0
0	0	1	0	0	0	0	0	0	0
1	0	0	0	0	0	1	0	0	0
2	0	0	0	0	0	0	1	0	0
0	0	1	0	0	0	0	0	0	0
1	0	0	0	0	0	1	0	0	0
2	0	0	0	0	0	0	1	0	0
0	0	1	0	0	0	0	0	0	0
1	0	0	0	0	0	1	0	0	0
2	0	0	0	0	0	0	1	0	0
0	0	1	0	0	0	0	0	0	0
1	0	0	0	1	0	0	0	0	0
0	0	1	0	0	0	0	0	0	0
1	0	0	0	1	0	0	0	0	0
F(ij)	1	6	1	4	1	3	3	1	0
P(ij)	0.125	0.75	0.125	0.5	0.125	0.375	0.75	0.25	0

Fig. 1.5. Building a Markov chain.

1.1.7 Stationary Distribution of a Finite Markov Chain

Definition 1.26. *A state i is said to have period d if $P_{ii}^{(n)} = 0$ whenever n is not divisible by d, and d is the largest integer with this property. A state with period 1 is said to be aperiodic.*

Example 1.27. Consider the transition probability matrix

$$P = \begin{pmatrix} 0 & 1 \\ 1 & 0 \end{pmatrix}.$$

We note that

$$P^{(n)} = \begin{pmatrix} 0 & 1 \\ 1 & 0 \end{pmatrix}^n = \frac{1}{2} \begin{pmatrix} 1+(-1)^n & 1+(-1)^{n+1} \\ 1+(-1)^{n+1} & 1+(-1)^n \end{pmatrix}.$$

We note that $P_{00}^{(2n+1)} = P_{11}^{(2n+1)} = 0$, so both States 0 and 1 have a period of 2.

Definition 1.28. *State i is said to be positive recurrent if it is recurrent and starting in state i the expected time until the process returns to state i is finite.*

Definition 1.29. *A state is said to be egordic if it is positive recurrent and aperiodic.*

We recall the example of the marketing problem with $\mathbf{X}^{(0)} = (1,0)^t$. We observe that

$$\mathbf{X}^{(1)} = P\mathbf{X}^{(0)} = \begin{pmatrix} 0.7 & 0.4 \\ 0.3 & 0.6 \end{pmatrix} (1,0)^T = (0.7, 0.3)^T,$$

$$\mathbf{X}^{(2)} = P^2\mathbf{X}^{(0)} = \begin{pmatrix} 0.61 & 0.52 \\ 0.39 & 0.48 \end{pmatrix} (1,0)^T = (0.61, 0.39)^T,$$

$$\mathbf{X}^{(4)} = P^4\mathbf{X}^{(0)} = \begin{pmatrix} 0.5749 & 0.5668 \\ 0.4251 & 0.4332 \end{pmatrix} (1,0)^T = (0.5749, 0.4251)^T,$$

$$\mathbf{X}^{(8)} = P^8\mathbf{X}^{(0)} = \begin{pmatrix} 0.5715 & 0.5714 \\ 0.4285 & 0.4286 \end{pmatrix} (1,0)^T = (0.5715, 0.4285)^T,$$

$$\mathbf{X}^{(16)} = P^{16}\mathbf{X}^{(0)} = \begin{pmatrix} 0.5714 & 0.5174 \\ 0.4286 & 0.4286 \end{pmatrix} (1,0)^T = (0.5714, 0.4286)^T.$$

It seems that

$$\lim_{n \to \infty} \mathbf{X}^{(n)} = (0.57\ldots, 0.42\ldots)^T.$$

In fact this limit exists and is independent of $\mathbf{X}^{(0)}$! It means that in the long run, the probability that a consumer belongs to Wellcome (Park'n) is given by 0.57 (0.42).

We note that $\mathbf{X}^{(n)} = P\mathbf{X}^{(n-1)}$ therefore if we let

$$\lim_{n\to\infty} \mathbf{X}^{(n)} = \pi$$

then

$$\pi = \lim_{n\to\infty} \mathbf{X}^{(n)} = \lim_{n\to\infty} P\mathbf{X}^{(n-1)} = P\pi.$$

We have the following definition

Definition 1.30. *A vector*

$$\pi = (\pi_0, \pi_1, \ldots, \pi_{k-1})^t$$

is said to be a stationary distribution of a finite Markov chain if it satisfies:

(i)

$$\pi_i \geq 0 \quad \text{and} \quad \sum_{i=0}^{k-1} \pi_i = 1.$$

(ii)

$$P\pi = \pi, \quad \text{i.e.} \sum_{j=0}^{k-1} P_{ij}\pi_j = \pi_i.$$

Proposition 1.31. *For any irreducible and aperiodic Markov chain having k states, there exists at least one stationary distribution.*

Proposition 1.32. *For any irreducible and aperiodic Markov chain having k states, for any initial distribution $\mathbf{X}^{(0)}$*

$$\lim_{n\to\infty} ||\mathbf{X}^{(n)} - \pi|| = \lim_{n\to\infty} ||P^n \mathbf{X}^{(0)} - \pi|| = 0.$$

where π is a stationary distribution for the transition matrix P.

Proposition 1.33. *The stationary distribution π in Proposition 1.32 is unique.*

There are a number of popular vector norms $||.||$. In the following, we introduce three of them.

Definition 1.34. *The $\mathbf{v}$ be a vector in R^n, then we have L_1-norm, L_∞-norm and 2-norm defined respectively by*

$$||\mathbf{v}||_1 = \sum_{i=1}^{n} |v_i|,$$

$$||\mathbf{v}||_\infty = \max_i \{|v_i|\},$$

and

$$||\mathbf{v}||_2 = \sqrt{\sum_{i=1}^{n} |v_i|^2}.$$

1.1.8 Applications of the Stationary Distribution

Recall the marketing problem again. The transition matrix is given by

$$P = \begin{pmatrix} 1 - \alpha & \beta \\ \alpha & 1 - \beta \end{pmatrix}.$$

To solve for the stationary distribution (π_0, π_1), we consider the following linear system of equations

$$\begin{cases} (1 - \alpha)\pi_0 + \beta\pi_1 & = \pi_0 \\ \alpha\pi_0 + (1 - \beta)\pi_1 & = \pi_1 \\ \pi_0 + \pi_1 & = 1. \end{cases}$$

Solving the linear system of equations we have

$$\begin{cases} \pi_0 = \beta(\alpha + \beta)^{-1} \\ \pi_1 = \alpha(\alpha + \beta)^{-1}. \end{cases}$$

Therefore in the long run, the market shares of Wellcome and Park'n are respectively

$$\frac{\beta}{(\alpha + \beta)} \quad \text{and} \quad \frac{\alpha}{(\alpha + \beta)}.$$

1.2 Continuous Time Markov Chain Process

In the previous section, we have discussed discrete time Markov chain processes. In many situations, a change of state does not occur at a fixed discrete time. In fact, the duration of a system state can be a continuous random variable. In our context, we are going to model queueing systems and re-manufacturing systems by continuous time Markov process. Here we first give the definition for a Poisson process. We then give some important properties of the Poisson process.

A process is called a Poisson process if

(A1) the probability of occurrence of one event in the time interval $(t, t + \delta t)$ is $\lambda\delta t + o(\delta t)$. Here λ is a positive constant and $o(\delta t)$ is such that

$$\lim_{\delta t \to 0} \frac{o(\delta t)}{\delta t} = 0.$$

(A2) the probability of occurrence of no event in the time interval $(t, t + \delta t)$ is $1 - \lambda\delta t + o(\delta t)$.

(A3) the probability of occurrences of more than one event is $o(\delta t)$.

Here an "event" can be an arrival of a bus or a departure of customer. From the above assumptions, one can derive the well-known Poisson distribution.

We define $P_n(t)$ be the probability that n events occurred in the time interval $[0, t]$. Assuming that that $P_n(t)$ is differentiable, then we can get a relationship between $P_n(t)$ and $P_{n-1}(t)$ as follows:

$$P_n(t + \delta t) = P_n(t) \cdot (1 - \lambda \delta t - o(\delta t)) + P_{n-1}(t) \cdot (\lambda \delta t + o(\delta t)) + o(\delta t).$$

Rearranging the terms we get

$$\frac{P_n(t + \delta t) - P_n(t)}{\delta t} = -\lambda P_n(t) + \lambda P_{n-1}(t) + (P_{n-1}(t) + P_n(t)) \frac{o(\delta t)}{\delta t}.$$

Let δt goes to zero, we have

$$\lim_{\delta t \to 0} \frac{P_n(t + \delta t) - P_n(t)}{\delta t} = -\lambda P_n(t) + \lambda P_{n-1}(t) + \lim_{\delta t \to 0} (P_{n-1}(t) + P_n(t)) \frac{o(\delta t)}{\delta t}.$$

Hence we have the differential-difference equation:

$$\frac{dP_n(t)}{dt} = -\lambda P_n(t) + \lambda P_{n-1}(t) + 0, \quad n = 0, 1, 2, \ldots.$$

Since $P_{-1}(t) = 0$, we have the *initial value problem* for $P_0(t)$ as follows:

$$\frac{dP_0(t)}{dt} = -\lambda P_0(t) \quad \text{with} \quad P_0(0) = 1.$$

The probability $P_0(0)$ is the probability that no event occurred in the time interval $[0, 0]$, so it must be one. Solving the separable ordinary differential equation for $P_0(t)$ we get

$$P_0(t) = e^{-\lambda t}$$

which is the probability that no event occurred in the time interval $[0, t]$. Thus

$$1 - P_0(t) = 1 - e^{-\lambda t}$$

is the probability that at least one event occurred in the time interval $[0, t]$. Therefore the probability density function $f(t)$ for the waiting time of the first event to occur is given by the well-known exponential distribution

$$f(t) = \frac{d(1 - e^{-\lambda t})}{dt} = \lambda e^{-\lambda t}, \quad t \geq 0.$$

We note that

$$\begin{cases} \frac{dP_n(t)}{dt} = -\lambda P_n(t) + \lambda P_{n-1}(t), \quad n = 1, 2, \ldots \\ P_0(t) = e^{-\lambda t}, \\ P_n(0) = 0 \quad n = 1, 2, \ldots. \end{cases}$$

Solving the above differential-difference equations, we get

$$P_n(t) = \frac{(\lambda t)^n}{n!} e^{-\lambda t}.$$

Finally, we present the important relationships among the Poisson process, Poisson distribution and the exponential distribution [52].

Proposition 1.35. *The following statements (B1),(B2), and (B3) are equivalent.*
(B1) The arrival process is a Poisson process with mean rate λ.
(B2) Let $N(t)$ be the number of arrivals in the time interval $[0,t]$ then

$$P(N(t) = n) = \frac{(\lambda t)^n e^{-\lambda t}}{n!} \quad n = 0,1,2,\ldots.$$

(B3) The inter-arrival time follows the exponential distribution with mean λ^{-1}.

1.2.1 A Continuous Two-state Markov Chain

Consider a one-server queueing system which has two possible states: 0 (idle) and 1 (busy). Assuming that the arrival process of the customers is a Poisson process with mean rate λ and the service time of the server follows the exponential distribution with mean rate μ. Let $P_0(t)$ be the probability that the server is idle at time t and $P_1(t)$ be the probability that the server is busy at time t. Using a similar argument as in the derivation of a Poisson process, we have

$$\begin{cases} P_0(t + \delta t) = (1 - \lambda \delta t - o(\delta t))P_0(t) + (\mu \delta t + o(\delta t))P_1(t) + o(\delta t) \\ P_1(t + \delta t) = (1 - \mu \delta t - o(\delta t))P_1(t) + (\lambda \delta t + o(\delta t))P_0(t) + o(\delta t). \end{cases}$$

Rearranging the terms, one gets

$$\begin{cases} \dfrac{P_0(t + \delta t) - P_0(t)}{\delta t} = -\lambda P_0(t) + \mu P_1(t) + (P_1(t) - P_0(t))\dfrac{o(\delta t)}{\delta t} \\ \dfrac{P_1(t + \delta t) - P_1(t)}{\delta t} = \lambda P_0(t) - \mu P_1(t) + (P_0(t) - P_1(t))\dfrac{o(\delta t)}{\delta t}. \end{cases}$$

Letting δt goes to zero, we get

$$\begin{cases} \dfrac{dP_0(t)}{dt} = -\lambda P_0(t) + \mu P_1(t) \\ \dfrac{dP_1(t)}{dt} = \lambda P_0(t) - \mu P_1(t). \end{cases}$$

Solving the above differential equations, we have

$$P_1(t) = \frac{1}{\lambda + \mu}(\mu e^{-(\lambda + \mu)t} + \lambda)$$

and
$$P_0(t) = \frac{1}{\lambda + \mu}(\mu - \mu e^{-(\lambda+\mu)t}).$$

We note that the steady state probabilities are given by
$$\lim_{t\to\infty} P_0(t) = \frac{\mu}{\lambda + \mu}$$

and
$$\lim_{t\to\infty} P_1(t) = \frac{\lambda}{\lambda + \mu}.$$

In fact, the steady state probability distribution can be obtained without solving the differential equations. We write the system of differential equations in matrix form:
$$\begin{pmatrix} \frac{dP_0(t)}{dt} \\ \frac{dP_1(t)}{dt} \end{pmatrix} = \begin{pmatrix} -\lambda & \mu \\ \lambda & -\mu \end{pmatrix} \begin{pmatrix} P_0(t) \\ P_1(t) \end{pmatrix}.$$

Since in steady state, $P_0(t) = p_0$ and $P_1(t) = p_1$ are constants and independent of t, we have
$$\frac{dp_0(t)}{dt} = \frac{dp_1(t)}{dt} = 0.$$

The steady state probabilities will be the solution of the following linear system:
$$\begin{pmatrix} -\lambda & \mu \\ \lambda & -\mu \end{pmatrix} \begin{pmatrix} p_0 \\ p_1 \end{pmatrix} = \begin{pmatrix} 0 \\ 0 \end{pmatrix}$$

subject to $p_0 + p_1 = 1$.

In fact, very often we are interested in obtaining the steady state probability distribution of the Markov chain. Because a lot of system performance such as expected number of customers, average waiting time can be written in terms of the steady state probability distribution, see for instance [48, 49, 50, 52]. We will also apply the concept of steady state probability distribution in the upcoming chapters. When the number of states is large, solving the steady state probability distribution will be time consuming. Iterative methods are popular approaches for solving large scale Markov chain problem.

1.3 Iterative Methods for Solving Linear Systems

In this section, we introduce some classical iterative methods for solving large linear systems. For more detail introduction to iterative methods, we refer reader to books by Bini et al. [21], Kincaid and Cheney [130], Golub and van Loan [101] and Saad [181].

1.3.1 Some Results on Matrix Theory

We begin our discussion by some more useful results in matrix theory and their proofs can be found in [112, 101, 130]. The first results is a useful formula for solving linear systems.

Proposition 1.36. *(Sherman-Morrison-Woodbury Formula) Let M be an non-singular $n \times n$ matrix, $\mathbf{u}$ and $\mathbf{v}$ be two $n \times k$ $(l \le n)$ matrices such that the matrix $(I_l + \mathbf{v}^T M \mathbf{u})$ is non-singular. Then we have*

$$\left(M + \mathbf{u}\mathbf{v}^T\right)^{-1} = M^{-1} - M^{-1}\mathbf{u}\left(I_l + \mathbf{v}^T M^{-1}\mathbf{u}\right)^{-1}\mathbf{v}^T M^{-1}.$$

The second result is on the eigenvalue of non-negative and irreducible square matrix.

Proposition 1.37. *(Perron-Frobenius Theorem) Let A be a non-negative and irreducible square matrix of order m. Then we have*
(i) A has a positive real eigenvalue λ which is equal to its spectral radius, i.e., $\lambda = \max_k |\lambda_k(A)|$ where $\lambda_k(A)$ denotes the k-th eigenvalue of A.
(ii) There corresponds an eigenvector $\mathbf{z}$ with all its entries being real and positive, such that $A\mathbf{z} = \lambda\mathbf{z}$.
(iii) λ is a simple eigenvalue of A.

The last result is on matrix norms. There are many matrix norms $||.||_M$ one can use. In the following, we introduce the definition of a matrix norm $||.||_{M_V}$ induced by a vector norm $||.||_V$.

Definition 1.38. *Given a vector $||.||_V$ in R^n, the matrix norm $||A||_{M_V}$ for an $n \times n$ matrix A induced by the vector norm is defined as*

$$||A||_{M_V} = \sup\{||A\mathbf{x}||_V : \mathbf{x} \in R^n \text{ and } ||\mathbf{x}||_V = 1\}.$$

In the following proposition, we introduce three popular matrix norms.

Proposition 1.39. *Let A be an $n \times n$ real matrix, then it can be shown that the matrix 1-norm, matrix ∞-norm and matrix 2-norm induced by $||.||_1$, $||.||_\infty$ and $||.||_2$ respectively by*

$$||A||_1 = \max_j \{\sum_{i=1}^n |A_{ij}|\},$$

$$||A||_\infty = \max_i \{\sum_{j=1}^n |A_{ij}|\},$$

and

$$||A||_2 = \sqrt{\lambda_{max}(AA^T)}.$$

Another other popular matrix norm is the Frobenius norm.

Definition 1.40. *The Frobenius norm of a square matrix A is defined as*

$$\|A\|_F = \sqrt{\sum_{i=1}^{n}\sum_{j=1}^{n} A_{ij}^2}.$$

1.3.2 Splitting of a Matrix

We begin with the concept of splitting a matrix. If we are to solve

$$A\mathbf{x} = \begin{pmatrix} \frac{1}{2} & \frac{1}{3} & 0 \\ \frac{1}{3} & 1 & \frac{1}{3} \\ 0 & \frac{1}{3} & \frac{1}{2} \end{pmatrix}\begin{pmatrix} x_1 \\ x_2 \\ x_3 \end{pmatrix} = \begin{pmatrix} 5 \\ 10 \\ 5 \end{pmatrix} = \mathbf{b}.$$

There are many ways to split the matrix A into two parts and develop iterative methods for solving the linear system.

There are at least three different ways of splitting the matrix A:

$$A = \begin{pmatrix} 1 & 0 & 0 \\ 0 & 1 & 0 \\ 0 & 0 & 1 \end{pmatrix} + \begin{pmatrix} \frac{-1}{2} & \frac{1}{3} & 0 \\ \frac{1}{3} & 0 & \frac{1}{3} \\ 0 & \frac{1}{3} & -\frac{1}{2} \end{pmatrix} \quad \text{(case 1)}$$

$$= \begin{pmatrix} \frac{1}{2} & 0 & 0 \\ 0 & 1 & 0 \\ 0 & 0 & \frac{1}{2} \end{pmatrix} + \begin{pmatrix} 0 & \frac{1}{3} & 0 \\ \frac{1}{3} & 0 & \frac{1}{3} \\ 0 & \frac{1}{3} & 0 \end{pmatrix} \quad \text{(case 2)}$$

$$= \begin{pmatrix} \frac{1}{2} & 0 & 0 \\ \frac{1}{3} & 1 & 0 \\ 0 & \frac{1}{3} & \frac{1}{2} \end{pmatrix} + \begin{pmatrix} 0 & \frac{1}{3} & 0 \\ 0 & 0 & \frac{1}{3} \\ 0 & 0 & 0 \end{pmatrix} \quad \text{(case 3)}$$

$$= S + (A - S)$$

Now

$$A\mathbf{x} = (S + (A - S))\mathbf{x} = \mathbf{b}$$

and therefore

$$S\mathbf{x} + (A - S)\mathbf{x} = \mathbf{b}$$

Hence we may write

$$\mathbf{x} = S^{-1}\mathbf{b} - S^{-1}(A - S)\mathbf{x}$$

where we assume that S^{-1} exists. Then given an initial guess $\mathbf{x}^{(0)}$ of the solution of $A\mathbf{x} = \mathbf{b}$, one may consider the following iterative scheme:

$$\mathbf{x}^{(k+1)} = S^{-1}\mathbf{b} - S^{-1}(A - S)\mathbf{x}^{(k)}. \tag{1.6}$$

Clearly if $\mathbf{x}^{(k)} \to \mathbf{x}$ as $k \to \infty$ then we have $\mathbf{x} = A^{-1}\mathbf{b}$. We note that (1.6) converges if and only if there is a matrix norm $\|.\|_M$ such that

$$\|S^{-1}(A - S)\|_M < 1.$$

This is because for any square matrix B, we have

$$(I - B)(I + B + B^2 + \ldots + B^n) = I - B^{n+1}$$

and

$$\sum_{k=0}^{\infty} B^k = (I - B)^{-1} \quad \text{if} \quad \lim_{n \to \infty} B^n = 0.$$

If there exists a matrix norm $||.|_M$ such that $||B||_M < 1$ then

$$\lim_{n \to \infty} ||B^n||_M \leq \lim_{n \to \infty} ||B||_M^n = 0$$

and we have

$$\lim_{n \to \infty} B^n = 0.$$

Therefore we have the following proposition.

Proposition 1.41. *If*

$$||S^{-1}(A - S)||_M < 1$$

then the iterative scheme converges to the solution of $Ax = b$.

1.3.3 Classical Iterative Methods

Throughout this section, we let A be the matrix to be split and $\mathbf{b}$ be the right hand side vector. We use $\mathbf{x}^{(0)} = (0, 0, 0)^T$ as the initial guess.

Case 1: $S = \begin{pmatrix} 1\ 0\ 0 \\ 0\ 1\ 0 \\ 0\ 0\ 1 \end{pmatrix}$.

$$\mathbf{x}^{(k+1)} = \mathbf{b} - (A - I)\mathbf{x}^{(k)}$$

$$= \begin{pmatrix} 5 \\ 10 \\ 5 \end{pmatrix} - \begin{pmatrix} -\frac{1}{2}\ \frac{1}{3}\ 0 \\ \frac{1}{3}\ 0\ \frac{1}{3} \\ 0\ \frac{1}{3}\ -\frac{1}{2} \end{pmatrix} \mathbf{x}^{(k)}$$

$$\mathbf{x}^{(1)} = (5\ 10\ 5)^T$$
$$\mathbf{x}^{(2)} = (4.1667\quad 6.6667\quad 4.1667)^T$$
$$\mathbf{x}^{(3)} = (4.8611\quad 7.2222\quad 4.8611)^T$$
$$\mathbf{x}^{(4)} = (5.0231\quad 6.7593\quad 5.0231)^T$$

$$\vdots$$

$$\mathbf{x}^{(30)} = (5.9983\quad 6.0014\quad 5.9983)^T.$$

When $S = I$, this is called the *Richardson method*.

Case 2: $S = \begin{pmatrix} \frac{1}{2} & 0 & 0 \\ 0 & 1 & 0 \\ 0 & 0 & \frac{1}{2} \end{pmatrix}$

Therefore

$$\mathbf{x}^{(k+1)} = S^{-1}\mathbf{b} - S^{-1}(A-S)\mathbf{x}^{(k)}$$

$$= \begin{pmatrix} 10 \\ 10 \\ 10 \end{pmatrix} - \begin{pmatrix} \frac{1}{2} & 0 & 0 \\ 0 & 1 & 0 \\ 0 & 0 & \frac{1}{2} \end{pmatrix}^{-1} \begin{pmatrix} 0 & \frac{1}{3} & 0 \\ \frac{1}{3} & 0 & \frac{1}{3} \\ 0 & \frac{1}{3} & 0 \end{pmatrix} \mathbf{x}^{(k)}$$

$$= (10\ 10\ 10)^T - \begin{pmatrix} 0 & \frac{2}{3} & 0 \\ \frac{1}{3} & 0 & \frac{1}{3} \\ 0 & \frac{2}{3} & 0 \end{pmatrix} \mathbf{x}^{(k)}$$

$$\mathbf{x}^{(1)} = (10\ 10\ 10)^T$$
$$\mathbf{x}^{(2)} = (3.3333 \quad 3.3333 \quad 3.3333)^T$$
$$\mathbf{x}^{(3)} = (7.7778 \quad 7.7778 \quad 7.7778)^T$$

$$\vdots$$

$$\mathbf{x}^{(30)} = (6.0000 \quad 6.0000 \quad 6.0000)^T.$$

When $S = \text{Diag}(a_{11}, \cdots, a_{nn})$. This is called the *Jacobi method*.

Case 3: $S = \begin{pmatrix} \frac{1}{2} & 0 & 0 \\ \frac{1}{3} & 1 & 0 \\ 0 & \frac{1}{3} & \frac{1}{2} \end{pmatrix}$

$$\mathbf{x}^{(k+1)} = S^{-1}\mathbf{b} - S^{-1}(A-S)\mathbf{x}^{(k)}$$

$$= \begin{pmatrix} 10 \\ \frac{20}{3} \\ \frac{50}{9} \end{pmatrix} - \begin{pmatrix} \frac{1}{2} & 0 & 0 \\ \frac{1}{3} & 1 & 0 \\ 0 & \frac{1}{3} & \frac{1}{2} \end{pmatrix}^{-1} \begin{pmatrix} 0 & \frac{1}{3} & 0 \\ 0 & 0 & \frac{1}{3} \\ 0 & 0 & 0 \end{pmatrix} \mathbf{x}^{(k)}$$

$$\mathbf{x}^{(1)} = (10\ \frac{20}{3}\ \frac{50}{9})^T$$
$$\mathbf{x}^{(2)} = (5.5556 \quad 6.2963 \quad 5.8025)^T$$
$$\mathbf{x}^{(3)} = (5.8025 \quad 6.1317 \quad 5.9122)^T$$
$$\mathbf{x}^{(4)} = (5.9122 \quad 6.0585 \quad 5.9610)^T$$

$$\vdots$$

$$\mathbf{x}^{(14)} = (6.0000 \quad 6.0000 \quad 6.0000)^T.$$

When S is the lower triangular part of the matrix A. This method is called the *Gauss-Seidel method*.

Proposition 1.42. *If A is diagonally dominant then*

$$||D^{-1}(A - D)||_\infty < 1$$

and the Jacobi method converges to the solution of $Ax = \mathbf{b}$.

1.3.4 Spectral Radius

Definition 1.43. *Given an $n \times n$ square matrix A the spectral radius of A is defined as*

$$\rho(A) = \max\{|\lambda| : \det(A - \lambda I) = 0\}$$

or in other words if $\lambda_1, \lambda_2, \cdots, \lambda_n$ are the eigenvalues of A then

$$\rho(A) = \max_i\{|\lambda_i|\}.$$

Example 1.44.

$$A = \begin{pmatrix} 0 & -1 \\ 1 & 0 \end{pmatrix}$$

then the eigenvalues of A are $\pm i$ and $|i| = |-i| = 1$. Therefore $\rho(A) = 1$ in this case.

Proposition 1.45. *For any square matrix A, $\rho(A) = \inf\limits_{||\cdot||_M} ||A||_M$.*

Remark 1.46. If $\rho(A) < 1$ then there exists a matrix norm $||.||_M$ such that $||A||_M < 1$.

Using the remark, one can show the following proposition.

Proposition 1.47. *The iterative scheme*

$$x^{(k)} = G\mathbf{x}^{(k-1)} + \mathbf{c}$$

converges to

$$(I - G)^{-1}\mathbf{c}$$

for any starting vectors $\mathbf{x}^{(0)}$ and $\mathbf{c}$ if and only if $\rho(G) < 1$.

Proposition 1.48. *The iterative scheme*

$$\mathbf{x}^{(k+1)} = S^{-1}\mathbf{b} - S^{-1}(A - S)\mathbf{x}^{(k)} = (I - S^{-1}A)\mathbf{x}^{(k)} + S^{-1}\mathbf{b}$$

converges to $A^{-1}\mathbf{b}$ if and only if $\rho(I - S^{-1}A) < 1$.

Proof. Take $G = I - S^{-1}A$ and $\mathbf{c} = S^{-1}\mathbf{b}$.

Definition 1.49. *An $n \times n$ matrix B is said to be strictly diagonal dominant if*

$$|B_{ii}| > \sum_{j=1, j \neq i}^{n} |B_{ij}| \quad \text{for} \quad i = 1, 2, \ldots, n$$

Proposition 1.50. *If A is strictly diagonally dominant then the Gauss-Seidel method converges for any starting $\mathbf{x}^{(0)}$.*

Proof. Let S be the lower triangular part of A. From Proposition 1.48 above, we only need to show

$$\rho(I - S^{-1}A) < 1.$$

Let λ be an eigenvalue of $(I - S^{-1}A)$ and $\mathbf{x}$ be its corresponding eigenvector such that

$$\|\mathbf{x}\|_\infty = 1.$$

We want to show

$$|\lambda| < 1.$$

We note that

$$(I - S^{-1}A)\mathbf{x} = \lambda \mathbf{x}$$

and therefore

$$\begin{pmatrix} 0 & -a_{12} & \cdots & -a_{1n} \\ \vdots & 0 & & \\ \vdots & & \ddots & -a_{n-1n} \\ 0 & \cdots & & 0 \end{pmatrix} \begin{pmatrix} x_1 \\ x_2 \\ \vdots \\ x_n \end{pmatrix} = \begin{pmatrix} a_{11} & 0 & \cdots & 0 \\ a_{21} & a_{22} & \ddots & \vdots \\ \vdots & & \ddots & 0 \\ a_{n1} & \cdots & \cdots & a_{nn} \end{pmatrix} \begin{pmatrix} \lambda x_1 \\ \lambda x_2 \\ \vdots \\ \lambda x_n \end{pmatrix}.$$

Therefore we have

$$-\sum_{j=i+1}^{n} a_{ij}x_j = \lambda \sum_{j=1}^{i} a_{ij}x_j \quad \text{for } i = 1, \cdots, n-1.$$

Since $\|x\|_\infty = 1$, there exists i such that

$$|x_i| = 1 \geq |x_j|.$$

For this i we have

$$|\lambda||a_{ii}| = |\lambda a_{ii}x_i| \leq \sum_{j=i+1}^{n} |a_{ij}| + |\lambda| \sum_{j=1}^{i-1} |a_{ij}|$$

and therefore

$$|\lambda| \leq \sum_{j=i+1}^{n} |a_{ij}| \left/ \left(|a_{ii}| - \sum_{j=1}^{i-1} |a_{ij}| \right) \right. < 1$$

1.3.5 Successive Over-Relaxation (SOR) Method

In solving $A\mathbf{x} = \mathbf{b}$, one may split A as follows:

$$A = \underbrace{L + wD} + (1 - w)D + U$$

where L is the strictly lower triangular part; D is the diagonal part and U is the strictly upper triangular part.

Example 1.51.

$$\begin{pmatrix} 2 & 1 & 0 \\ 1 & 2 & 1 \\ 0 & 1 & 2 \end{pmatrix} = \underbrace{\begin{pmatrix} 0 & 0 & 0 \\ 1 & 0 & 0 \\ 0 & 1 & 0 \end{pmatrix}}_{L} + w \underbrace{\begin{pmatrix} 2 & 0 & 0 \\ 0 & 2 & 0 \\ 0 & 0 & 2 \end{pmatrix}}_{D} + (1 - w) \underbrace{\begin{pmatrix} 2 & 0 & 0 \\ 0 & 2 & 0 \\ 0 & 0 & 2 \end{pmatrix}}_{D} + \underbrace{\begin{pmatrix} 0 & 1 & 0 \\ 0 & 0 & 1 \\ 0 & 0 & 0 \end{pmatrix}}_{U}$$

One may consider the iterative scheme with $S = L + wD$ as follows:

$$\mathbf{x}_{n+1} = S^{-1}\mathbf{b} + S^{-1}(S - A)\mathbf{x}_n = S^{-1}\mathbf{b} + (I - S^{-1}A)\mathbf{x}_n.$$

We remark that
$$I - S^{-1}A = I - (L + wD)^{-1}A.$$

Moreover, when $w = 1$, it is just the Gauss-Seidel method. This method is called the *SOR method*. It is clear that the method converges if and only if the iteration matrix has a spectral radius less than one.

Proposition 1.52. *The SOR method converges to the solution of $A\mathbf{x} = \mathbf{b}$ if and only if $\rho(I - (L + wD)^{-1}A) < 1$.*

1.3.6 Conjugate Gradient Method

Conjugate gradient (CG) methods are iterative methods for solving linear system of equations $A\mathbf{x} = \mathbf{b}$ where A is symmetric positive definite [11, 101]. This method was first discussed by Hestenes and Stiefel [109]. The motivation of the method is that it involves the process of minimizing quadratic functions such as
$$f(\mathbf{x}) = (A\mathbf{x} - \mathbf{b})^T(A\mathbf{x} - \mathbf{b}).$$

Here A is symmetric positive definite and this minimization usually takes place over a sequence of Krylov subspaces which is generated recursively by adding a new basic vector $A^k \mathbf{r}_0$ to those of the subspace V_{k-1} generated where

$$\mathbf{r}_0 = A\mathbf{x}_0 - \mathbf{b}$$

is the residue of the initial vector $\mathbf{x}_0$.

Usually, a sequence of conjugate orthogonal vectors is constructed from V_k so that CG methods would be more efficient. Computing these vectors can be done recursively which involves only a few vectors if A is self-adjoint with respect to the inner product. The CG methods are attractive since they can give the exact solution after in most n steps in exact arithmetic where n is the size of the matrix A. Hence it can also be regarded as a direct method in this sense. But in the presence of round off errors and finite precision, the number of iterations may be greater than n. Thus, CG methods can be seen as least square methods where the minimization takes place on a particular vector subspace, the Krylov space. When estimating the error of the current solution in each step, a matrix-vector multiplication is then needed. The CG methods are popular and their convergence rates can be improved by using suitable preconditioning techniques. Moreover, it is parameter free, the recursion involved are usually short in each iteration and the memory requirements and the execution time are acceptable for many practical problems.

The CG algorithm reads:

Given an initial guess $\mathbf{x}^0$, A, $\mathbf{b}$, Max, tol:

$\mathbf{r}^0 = \mathbf{b} - A\mathbf{x}^0$;

$\mathbf{v}^0 = \mathbf{r}^0$;

For $k = 0$ to Max-1 do

If $||\mathbf{v}^k||_2 = 0$ then stop

$t_k = <\mathbf{r}^k, \mathbf{r}^k> / <\mathbf{v}^k, A\mathbf{v}^k>$;

$\mathbf{x}^{k+1} = \mathbf{x}^k + t_k\mathbf{v}^k$;

$\mathbf{r}^{k+1} = \mathbf{r}^k - t_k A\mathbf{v}^k$;

If $||\mathbf{r}^{k+1}, \mathbf{r}^{k+1}||_2 < $ tol then stop

$\mathbf{v}^{k+1} = \mathbf{r}^{k+1} + <\mathbf{r}^{k+1}, \mathbf{r}^{k+1}> / <\mathbf{r}^k, \mathbf{r}^k> \mathbf{v}^k$;

end;

output $\mathbf{x}^{k+1}, ||\mathbf{r}^{k+1}||_2$.

Given a Hermitian, positive definite $n \times n$ matrix H_n, when the conjugate gradient method is applied to solving

$$H_n\mathbf{x} = \mathbf{b}$$

the convergence rate of this method depends on the spectrum of the matrix H_n, see also Golub and van Loan [101]. For example if the spectrum of H_n is contained in an interval, i.e. $\sigma(H_n) \subseteq [a, b]$, then the error in the i-th iteration is given by

$$\frac{||\mathbf{e}_i||}{||\mathbf{e}_0||} \leq 2(\frac{\sqrt{b} - \sqrt{a}}{\sqrt{b} + \sqrt{a}})^i,$$

i.e. the convergence rate is linear. Hence the approximate upper bound for the number of iterations required to make the relative error

$$\frac{||\mathbf{e}_i||}{||\mathbf{e}_0||} \leq \delta$$

is given by

$$\frac{1}{2}(\sqrt{\frac{b}{a}} - 1) \log(\frac{2}{\delta}) + 1.$$

Very often CG method is used with a matrix called preconditioner to accelerate its convergence rate. A good preconditioner C should satisfy the following conditions.

(i) The matrix C can be constructed easily;
(ii) Given right hand side vector $\mathbf{r}$, the linear system $C\mathbf{y} = \mathbf{r}$ can be solved efficiently; and
(iii) the spectrum (or singular values) of the preconditioned system $C^{-1}A$ should be clustered around one.

In the Preconditioned Conjugate Gradient (PCG) method, we solve the linear system

$$C^{-1}A\mathbf{x} = C^{-1}\mathbf{b}$$

instead of the original linear system

$$A\mathbf{x} = \mathbf{b}.$$

We expect the fast convergence rate of the PCG method can compensate much more than the extra cost in solving the preconditioner system $C\mathbf{y} = \mathbf{r}$ in each iteration step of the PCG method.

Apart from the approach of condition number, in fact, condition (iii) is also very commonly used in proving convergence rate. In the following we give the definition of clustering.

Definition 1.53. *We say that a sequence of matrices S_n of size n has a clustered spectrum around one if for all $\epsilon > 0$, there exist non-negative integers n_0 and n_1, such that for all $n > n_0$, at most n_1 eigenvalues of the matrix $S_n^* S_n - I_n$ have absolute values larger than ϵ.*

One sufficient condition for the matrix to have eigenvalues clustered around one is that

$$H_n = I_n + L_n,$$

where I_n is the $n \times n$ identity matrix and L_n is a low rank matrix ($\text{rank}(L_n)$ is bounded above and independent of the matrix size n).

Conjugate Gradient Squared Method

Given a real symmetric, positive definite matrix A of size $n \times n$, the CG method can be used to solve the linear system $A\mathbf{x} = \mathbf{b}$. But in general a non-singular matrix can be neither symmetric nor positive definite. In particular for the applications in queueing systems and re-manufacturing systems in Chapters 2 and 3. In this case, one may consider the normal equation of the original system. i.e.,

$$A^T A \mathbf{x} = A^T \mathbf{b}.$$

Here $A^T A$ is real symmetric and positive definite so that CG method could be applied, but the condition number would then be squared. Moreover, it also involves the matrix-vector multiplication of the form $A^T \mathbf{r}$. These will increase the computational cost. Thus in our context, we propose to employ a generalized CG algorithm, namely the Conjugate Gradient Squared (CGS) method, [193]. This method does not involve the matrix-vector multiplication of the form $A^T \mathbf{r}$.

The CGS algorithm reads:

Given an initial guess $\mathbf{x}^0$, A, $\mathbf{b}$, tol:

$\mathbf{x} = \mathbf{x}_0$;

$\mathbf{r} = \mathbf{b} - A\mathbf{x}$;

$\mathbf{r}' = \mathbf{s} = \mathbf{p} = \mathbf{r}$;

$\mathbf{w} = A\mathbf{p}$;

$\mu = {\mathbf{r}'}^T \mathbf{r}$;

repeat until $\mu < \text{tol}$;

$\gamma = \mu$;

$\alpha = \gamma / {\mathbf{r}'}^t \mathbf{r}$;

$\mathbf{q} = \mathbf{s} - \alpha \mathbf{w}$;

$\mathbf{d} = \mathbf{s} + \mathbf{q}$;

$\mathbf{w} = A\mathbf{d}$;

$\mathbf{x} = \mathbf{x} + \alpha \mathbf{d}$;

$\mathbf{r} = \mathbf{r} - \alpha\mathbf{w};$

otherwise

$\mu = \mathbf{r'}^T\mathbf{r};$

$\beta = \mu/\gamma;$

$\mathbf{s} = \mathbf{r} - \beta\mathbf{q};$

$\mathbf{p} = \mathbf{s} + \beta(\mathbf{q} + \beta\mathbf{p});$
end;

1.3.7 Toeplitz Matrices

We end this subsection by introducing a class of matrices, namely Toeplitz matrices. A Toepltiz matrix T is a matrix having constant diagonals, i.e.

$$T = \begin{pmatrix} t_0 & t_1 & t_2 & \cdots & t_{n-1} & t_n \\ t_{-1} & t_0 & t_1 & \cdots & \cdots & t_{n-1} \\ \vdots & \ddots & \ddots & \ddots & \ddots & \vdots \\ \vdots & \ddots & \ddots & \ddots & \ddots & \vdots \\ t_{-n+1} & \cdots & \cdots & \ddots & \ddots & t_1 \\ t_{-n} & t_{-n+1} & \cdots & \cdots & t_{-1} & t_0 \end{pmatrix}.$$

Toeplitz matrices and near-Toeplitz matrices have many applications in applied sciences and engineering such as the multi-channel least squares filtering in time series [171], signal and image processing problems [145]. A survey on the applications of Toeplitz systems can be found in Chan and Ng [46]. Application in solving queueing systems and re-manufacturing systems will be discussed in the Chapters 2 and 3.

In the above applications, solving a Toeplitz or near-Toeplitz system is the focus. Direct methods for solving Toeplitz systems based on the recursion formula are commonly used, see for instance, Trench [199]. For an $n \times n$ Toeplitz matrix T, these direct methods require $O(n^2)$ operations. Faster algorithms that require $O(n \log^2 n)$ operations have also been developed when the Toeplitz matrix is symmetric and positive definite.

An important subset of Toepltiz matrices is the class of circulant matrices. A circulant $n \times n$ matrix C is a Toeplitz matrix such that each column is a cyclic shift of the previous one, i.e.

$$C = \begin{pmatrix} c_0 & c_1 & \cdots & c_{n-1} & c_n \\ c_n & c_0 & c_1 & \cdots & c_{n-1} \\ \vdots & \ddots & \ddots & \ddots & \vdots \\ c_2 & \vdots & \ddots & \ddots & c_1 \\ c_1 & c_2 & \cdots & c_n & c_0 \end{pmatrix}. \tag{1.7}$$

Very often circulant matrices are used to approximate Toeplitz matrices in preconditioning or finding approximate solution. Because circulant matrices have the following nice property. It is well-known that a circulant matrix can be diagonalized by the discrete Fourier matrix F. More precisely,

$$FCF^* = D = \text{Diag}(d_0, d_1, \ldots, d_n)$$

where F is the discrete Fourier matrix with entries given by

$$F_{j,k} = \frac{1}{\sqrt{n}} e^{-\frac{(2jk\pi)i}{n}}, \quad j, k = 0, 1, \cdots, n-1,$$

and D is a diagonal matrix with elements being the eigenvalues of C, see for instance [82]. Here F^* is the conjugate transpose of F. The matrix-vector multiplication $F\mathbf{y}$ is called the Fast Fourier Transformation (FFT) of the column vector $\mathbf{y}$ and can be done in $O(n \log n)$ operations. Consider for a unit vector

$$\mathbf{e}_1 = (1, 0, \ldots, 0)^T,$$

we have

$$C\mathbf{e}_1 = (c_0, c_n, \ldots, c_1)^T$$

and

$$F\mathbf{e}_1 = \frac{1}{\sqrt{n}}(1, 1, \ldots, 1)^T$$

because the first column of F is a column vector with all entries being equal. Therefore

$$F(c_0, c_n, \ldots, c_1)^T = FC\mathbf{e}_1 = DF\mathbf{e}_1 = \frac{1}{\sqrt{n}}(d_0, d_1, \ldots, d_n)^T$$

and hence the eigenvectors of a circulant matrix C can be obtained by using the FFT in $O(n \log n)$ operations. Moreover, the solution of a circulant linear system can also be obtained in $O(n \log n)$ operations.

The FFT can be used in the Toeplitz matrix-vector multiplication. A Toeplitz matrix can be embedded in a circulant matrix as follows:

$$\tilde{C}(\mathbf{y}, \mathbf{0})^T \equiv \begin{pmatrix} T & S_1 \\ S_2 & T \end{pmatrix} \begin{pmatrix} \mathbf{y} \\ \mathbf{0} \end{pmatrix} = \begin{pmatrix} \mathbf{r} \\ \mathbf{b} \end{pmatrix}. \tag{1.8}$$

Here matrices S_1 and S_2 are such that $\tilde{C}$ is a circulant matrix. Then FFT can be applied to obtain $\mathbf{r} = T\mathbf{y}$ in $O(n \log n)$ operations.

1.4 Hidden Markov Models

Hidden Markov Models (HMMs) are widely used in bioinformatics [135], speech recognition [173] and many other areas [149]. In a HMM, there are two types of states: the observable states and the hidden states. In a HMM, there is no one-to-one correspondence between the hidden states and the observed symbols. It is therefore no longer possible to tell what hidden state the model is in which the observation symbol is generated just by looking at the observation symbol. A HMM is usually characterized by the following elements [173]:

- N, the number of hidden states in the model. Although the states are hidden, for many practical applications there is often some physical significance to the states. For instance, the hidden states represent the CpG island and the non-CpG island in the DNA sequence. We denote the individual states as
$$S = \{s_1, s_2, \cdots, s_N\},$$
and the state at the length t as Q_t.
- M, the number of distinct observation symbols per hidden state. The observation symbols correspond to the physical output of the system being modeled. For instance, A,C,G,T are the observation symbols in the DNA sequence. We denote the individual symbols as
$$V = \{v_1, v_2, \cdots, v_M\}$$
and the symbol at the length t as O_t.
- The state transition probability distribution $[A]_{ij} = \{a_{ij}\}$ where
$$a_{ij} = P(Q_{t+1} = s_j | Q_t = s_i), \quad 1 \le i, j \le N.$$
- The observation symbol probability distribution in hidden state j, $[B]_{jk} = \{b_j(v_k)\}$, where
$$b_j(v_k) = P(O_t = v_k | Q_t = s_j), \quad 1 \le j \le N, \quad 1 \le k \le M.$$
- The initial state distribution $\Pi = \{\pi_i\}$ where
$$\pi_i = P(Q_1 = s_i), \quad 1 \le i \le N.$$

Given appropriate values of N, M, A, B and Π, the HMM can be used as a generator to give an observation sequence
$$O = \{O_1 O_2 O_3 \cdots O_T\}$$
where T is the number of observations in the sequence. For simplicity, we use the compact notation
$$\Lambda = (A, B, \Pi)$$
to indicate the complete parameter set of the HMM. According to the above specification, the first order transition probability distribution among the hidden states is used. There are three key issues in HHMMs:

- **Problem 1:**
  Given the observation sequence $O = \{O_1 O_2 \cdots O_T\}$ and a HMM, how to efficiently compute the probability of the observation sequence ?
- **Problem 2:**
  Given the observation sequence $O = \{O_1 O_2 \cdots O_T\}$ and a HMM, how to choose a corresponding state sequence $Q = \{Q_1 Q_2 \cdots Q_T\}$ which is optimal in certain sense ?
- **Problem 3:** Given the observation sequence $O = \{O_1 O_2 \cdots O_T\}$, how to choose the model parameters in a HMM?

For **Problem 1**, a forward-backward dynamic programming procedure [14] is formulated to calculate the probability of the observation sequence efficiently.

For **Problem 2**, it is the one in which we attempt to uncover the hidden part of the model, i.e., to find the "correct" state sequence. In many practical situations, we use an optimality criteria to solve the problem as good as possible. The most widely used criterion is to find a single best state sequence, i.e., maximize the likelihood $P(Q|A, O)$. This is equivalent to maximizing $P(Q, O|A)$ since

$$P(Q|A, O) = \frac{P(Q, O|A)}{P(O|A)}.$$

Viterbi algorithm [204] is a dynamic programming technique for finding this single best state sequence

$$Q = \{Q_1, Q_2, \cdots, Q_T\}$$

for the given observation sequence

$$O = \{O_1, O_2, \cdots, O_T\}.$$

For **Problem 3**, we attempt to adjust the model parameters A such that $P(O|A)$ is maximized by using Expectation-Maximization (EM) algorithm. For a complete tutorial on hidden Markov model, we refer readers to the paper by Rabiner [173] and the book by MacDonald and Zucchini [149].

1.5 Markov Decison Process

Markov Decision Process (MDP) has been successfully applied in equipment maintenance, inventory control and many other areas in management science [4, 209]. In this section, we will briefly introduce the MDP, interested readers can also consult the books by Altman [4], Puterman [172] and White [208].

Similar to the case of Markov chain, MDP is a system that can move from one distinguished state to any other possible states. In each step, the decision maker has to take an action from a well-defined set of alternatives. This action affects the transition probabilities of the next move and incurs an immediate gain (or loss) and subsequent gain (or loss). The obvious problem that the decision maker facing is to determine a suitable plan of actions so that the overall gain is optimized. The process of MDP is summarized as follows:

(i) At time t, a certain state i of the Markov chain is observed.
(ii) After the observation of the state, an action, let us say k is taken from a set of possible decisions A_i. Different states may have different sets of decisions.
(iii) An immediate gain (or loss) $q_i^{(k)}$ is then incurred according to the current state i and the action k taken.
(iv) The transition probabilities $p_{ji}^{(k)}$ is then affected by the action k.
(v) When the time parameter t increases, transition occurs again and the above steps (i)-(iv) repeat.

A policy D is a rule of taking actions. It prescribes all the decisions that should be made throughout the process. Given the current state i, the value of an optimal policy $v_i(t)$ is defined as the total expected gain obtained with t decisions or transitions remained. For the case of one-period remaining, i.e. $t = 1$, the value of an optimal policy is given by

$$v_i(1) = \max_{k \in A_i}\{q_i^{(k)}\}. \tag{1.9}$$

Since there is only one-period remained, an action maximizing the immediate gain will be taken. For the case of two-period remaining, we have

$$v_i(2) = \max_{k \in A_i}\{q_i^{(k)} + \alpha \underbrace{\sum_j p_{ji}^{(k)} v_j(1)}_{\text{subsequent gain}}\} \tag{1.10}$$

where α is the discount factor. Since that the subsequent gain is associated with the transition probabilities which are affected by the actions, an optimal policy should consider both the immediate and subsequent gain. The model can be easily extended to a more general situation, the process having n transitions remained.

$$v_i(n) = \max_{k \in A_i}\{q_i^{(k)} + \alpha \underbrace{\sum_j p_{ji}^{(k)} v_j(n-1)}_{\text{subsequent gain}}\}. \tag{1.11}$$

From the above equation, the subsequent gain of $v_i(n)$ is defined as the expected value of $v_j(n-1)$. Since the number of transitions remained is countable or finite, the process is called the discounted finite horizon MDP. For the infinite horizon MDP, the value of an optimal policy can be expressed as

$$v_i = \max_{k \in A_i}\{q_i^{(k)} + \alpha \sum_j p_{ji}^{(k)} v_j\}. \tag{1.12}$$

The finite horizon MDP is a dynamic programming problem and the infinite horizon MDP can be transformed into a linear programming problem. Both of them can be solved easily by using EXCEL spreadsheet.

1.5.1 Stationary Policy

A stationary policy is a policy that the choice of alternative depends only on the state the system is in and is independent of n. For instance, a stationary policy D prescribes the action $D(i)$ when the current state is i. Define $\bar{D}$ as the associated one-step-removed policy, then the value of policy $w_i(D)$ is defined as

$$w_i(D) = q_i^{D(i)} + \alpha \sum_j p_{ji}^{D(i)} w_j(\bar{D}). \tag{1.13}$$

Given a Markov decision process with infinite horizon and discount factor α, $0 < \alpha < 1$, choose, for each i, an alternative k_i such that

$$\max_{k \in A_i} \{ q_i^{(k)} + \alpha \sum_j p_{ji}^{(k)} v_j \} = q_i^{(k_i)} + \alpha \sum_j p_{ji}^{(k_i)} v_j.$$

Define the stationary policy D by $D(i) = k_i$. Then for each i, $w_i(D) = v_i$, i.e. the stationary policy is an optimal policy.

2

Queueing Systems and the Web

In this chapter, we will first discuss some more Markovian queueing systems. The queueing system is a classical application of continuous Markov chain. We then present an important numerical algorithm based on computation of Markov chain for ranking the webpages in the Web. This is a modern applications of Markov though the numerical methods used are classical.

2.1 Markovian Queueing Systems

An important class of queueing networks is the Markovian queueing systems. The main assumptions of a Markovian queueing system are the Poisson arrival process and exponential service time. The one-server system discussed in the previous section is a queueing system without waiting space. This means when a customer arrives and finds the server is busy, the customer has to leave the system. In the following sections, we will introduce some more Markovian queueing systems. Queueing system is a classical application of continuous time Markov chain. We will further discuss its applications in re-manufacturing systems in Chapter 3. For more details about numerical solutions for queueing system and Markov chain, we refer the read to the books by Ching [52], Leonard [144], Neuts [159, 160] and Stewart [194].

2.1.1 An $M/M/1/n - 2$ Queueing System

Now let us consider a more general queueing system with customer arrival rate being λ. Suppose the system has one exponential servers with service rate being μ and there are $n - 2$ waiting spaces in the system. The queueing discipline is *First-come-first-served*. When an arrived customer finds the server is busy, then customer can still wait in the queue provided that there is a waiting space available. Otherwise, the customer has to leave the queueing system. To describe the queueing system, we use the number of customers in the queue to represent the state of the system. There are n states, namely

$0, 1, \ldots, n-1$. The Markov chain for the queueing system is given in Fig. 2.1. The number of customers in the system is used to represent the states in the Markov chain. Clearly it is an irreducible Markov chain.

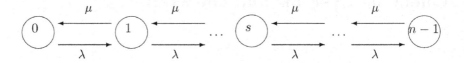

Fig. 2.1. The Markov chain for the one-queue system.

If we order the states of the system in increasing number of customers, it is not difficult to show that the generator matrix for this queueing system is given by the following $n \times n$ tri-diagonal matrix $A_1 = A_{(n,1,\lambda,\mu)}$ where

$$
A_1 = \begin{pmatrix}
\lambda & -\mu & & & & & & 0 \\
-\lambda & \lambda+\mu & -\mu & & & & & \\
& \ddots & \ddots & \ddots & & & & \\
& & -\lambda & \lambda+\mu & -\mu & & & \\
& & & -\lambda & \lambda+\mu & -\mu & & \\
& & & & \ddots & \ddots & \ddots & \\
& & & & & -\lambda & \lambda+\mu & -\mu \\
0 & & & & & & -\lambda & s\mu
\end{pmatrix}
\tag{2.1}
$$

and the underlying Markov chain is irreducible. The solution for the steady-state probability distribution can be shown to be

$$
\mathbf{p}_{(n,1,\lambda,\mu)}^T = (p_0, p_1, \ldots, p_{n-1})^T
\tag{2.2}
$$

where

$$
p_i = \alpha \prod_{k=1}^{i+1} \frac{\lambda}{\mu} \quad \text{and} \quad \alpha^{-1} = \sum_{i=0}^{n} p_i.
\tag{2.3}
$$

Here p_i is the probability that there are i customers in the queueing system in the steady state and α is the *normalization constant*.

Example 2.1. Consider a one-server system; the steady-state probability distribution is given by

$$
p_i = \frac{\rho^i(1-\rho)}{1-\rho^n} \quad \text{where} \quad \rho = \frac{\lambda}{\mu}.
$$

When the system has no limit on waiting space and $\rho < 1$, the steady-state probability becomes

$$\lim_{n \to \infty} p_i = \rho^i (1 - \rho).$$

The expected number of customers in the system is given by

$$
\begin{aligned}
L_c &= \sum_{i=0}^{\infty} i p_i \\
&= \sum_{i=0}^{\infty} i \rho^i (1 - \rho) \\
&= \frac{\rho(1 - \rho)}{(1 - \rho)^2} = \frac{\rho}{1 - \rho}.
\end{aligned}
$$

The expected number of customers waiting in the queue is given by

$$
\begin{aligned}
L_q &= \sum_{i=1}^{\infty} (i - 1) p_i \\
&= \sum_{i=1}^{\infty} (i - 1) \rho^i (1 - \rho) \\
&= \frac{\rho}{1 - \rho} - \rho.
\end{aligned}
$$

Moreover the expected number of customers in service is given by

$$
L_s = 0 \cdot p_0 + 1 \cdot \sum_{i=1}^{\infty} p_i = 1 - (1 - \rho) = \rho.
$$

2.1.2 An M/M/s/$n - s - 1$ Queueing System

Now let us consider a more general queueing system with customer arrival rate being λ. Suppose the system has s parallel and identical exponential servers with service rate being μ and there are $n - s - 1$ waiting spaces in the system. The queueing discipline is *First-come-first-served*. Again when a customer arrives and finds all the servers are busy, the customer can still wait in the queue provided that there is a waiting space available. Otherwise, the customer has to leave the system. To apply the continuous time Markov chain for model this queueing system, one has to obtain the waiting for one departure of customer when there are more than one customer (let us say k customers) in the queueing system. We need the following lemma

Lemma 2.2. *Suppose that* $X_1, X_2, \ldots, X_k$ *are independent, identical, exponential random variables with mean* μ^{-1}, *and consider the corresponding order statistics*

$$X_{(1)} \le X_{(2)} \le \cdots \le X_{(k)}.$$

Then $X_{(1)}$ *is again exponentially distributed with mean* $\frac{1}{k}$ *times the mean of the original random variables.*

Proof. We observe that

$$X_{(1)} = \min(X_1, X_2, \ldots, X_k).$$

$X_{(1)} > x$ if and only if all $X_i > x$ $(i = 1, 2, \ldots, k)$. Hence

$$\begin{aligned} P\{X_{(1)} > x\} &= P\{X_1 > x\}P\{X_2 > x\} \cdots P\{X_k > x\} \\ &= (e^{-\mu x})^k \\ &= e^{-k\mu x}. \end{aligned}$$

Again it is still exponentially distributed with mean $1/(k\mu)$. If we use the number of customers in the queue to represent the state of the system. There are n states, namely $0, 1, \ldots, n-1$. The Markov chain for the queueing system is given in Fig. 2.2. The number of customers in the system is used to represent the states in the Markov chain. Clearly it is an irreducible Markov chain.

Fig. 2.2. The Markov chain for the one-queue system.

If we order the states of the system in increasing number of customers, it is not difficult to show that the generator matrix for this queueing system is given by the following $n \times n$ tri-diagonal matrix $A_2 = A_{(n,s,\lambda,\mu)}$ where

$$A_2 = \begin{pmatrix} \lambda & -\mu & & & & & & 0 \\ -\lambda & \lambda + \mu & -2\mu & & & & & \\ & \ddots & \ddots & \ddots & & & & \\ & & -\lambda & \lambda + (s-1)\mu & -s\mu & & & \\ & & & -\lambda & \lambda + s\mu & -s\mu & & \\ & & & & \ddots & \ddots & \ddots & \\ & & & & & -\lambda & \lambda + s\mu & -s\mu \\ 0 & & & & & & -\lambda & s\mu \end{pmatrix} \quad (2.4)$$

and the underlying Markov chain is irreducible. The solution for the steady-state probability distribution can be shown to be

$$\mathbf{p}^T_{(n,s,\lambda,\mu)} = (p_0, p_1, \ldots, p_{n-1})^T \quad (2.5)$$

where

$$p_i = \alpha \prod_{k=1}^{i+1} \frac{\lambda}{\mu \min\{k, s\}}$$

and

$$\alpha^{-1} = \sum_{i=0}^{n} p_i.$$

Here p_i is the probability that there are i customers in the queueing system in steady state and α is the *normalization constant*.

2.1.3 The Two-Queue Free System

In this subsection, we introduce a higher dimensional queueing system. Suppose that there are two one-queue systems as discussed in Section 2.1.2. This queueing system consists of two independent queues with the number of identical servers and waiting spaces being s_i and $n_i - s_i - 1$ ($i = 1, 2$) respectively. It we let the arrival rate of customers in the queue i be λ_i and service rate of the servers be μ_i ($i = 1, 2$) then the states of the queueing system can be represented by the elements in the following set:

$$S = \{(i, j) | 0 \leq i \leq n_1, 0 \leq j \leq n_2\}$$

where (i, j) represents the state that there are i customers in queue 1 and j customers in queue 2. Thus this is a two-dimensional queueing model. If we order the states lexicographically, then the generator matrix can be shown to be the following $n_1 n_2 \times n_1 n_2$ matrix in *tensor product* form [44, 52]:

$$A_3 = I_{n_1} \otimes A_{(n_2, s_2, \lambda_2, \mu_2)} + A_{(n_1, s_1, \lambda_1, \mu_1)} \otimes I_{n_2}. \tag{2.6}$$

Here $\otimes$ is the Kronecker tensor product [101, 112]. The Kronecker tensor product of two matrices A and B of sizes $p \times q$ and $m \times n$ respectively is a $(pm) \times (qn)$ matrix given as follows:

$$A \otimes B = \begin{pmatrix} a_{11}B & \cdots & \cdots & a_{1q}B \\ a_{21}B & \cdots & \cdots & a_{2q}B \\ \vdots & \vdots & \vdots & \vdots \\ a_{p1}B & \cdots & \cdots & a_{pq}B \end{pmatrix}.$$

The Kronecker tensor product is a useful tool for representing generator matrices in many queueing systems and stochastic automata networks [44, 52, 138, 194]. For this two-queue free queueing system, it is also not difficult to show that the steady state probability distribution is given by the probability distribution vector

$$\mathbf{P}_{(n_1, s_1, \lambda_1, \mu_1)} \otimes \mathbf{P}_{(n_2, s_2, \lambda_2, \mu_2)}. \tag{2.7}$$

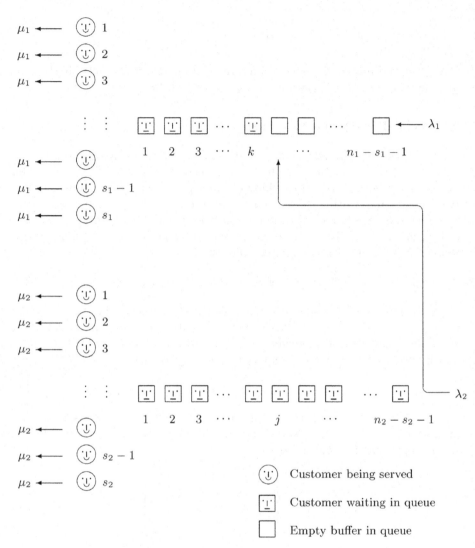

Fig. 2.3. The two-queue overflow system.

2.1.4 The Two-Queue Overflow System

Now let us add the following system dynamics to the two-queue free system discussed Section 2.1.3. In this queueing system, we allow overflow of customers from queue 2 to queue 1 whenever queue 2 is full and there is still waiting space in queue 1; see for instance Fig. 2.3 (Taken from [52]). This is called the two-queue overflow system; see Kaufman [44, 52, 136].

In this case, the generator matrix is given by the following matrix:

$$A_4 = I_{n_1} \otimes A_{(n_2,s_2,\lambda_2,\mu_2)} + A_{(n_1,s_1,\lambda_1,\mu_1)} \otimes I_{n_2} + R \otimes \mathbf{e_{n_2}}{}^t\mathbf{e_{n_2}}. \quad (2.8)$$

Here $\mathbf{e_{n_2}}$ is the unit vector $(0, 0, \ldots, 0, 1)$ and

$$R = \begin{pmatrix} \lambda_2 & & & & 0 \\ -\lambda_2 & \lambda_2 & & & \\ & -\lambda_2 & \ddots & & \\ & & \ddots & \lambda_2 & \\ 0 & & & -\lambda_2 & 0 \end{pmatrix}. \quad (2.9)$$

In fact

$$A_4 = A_3 + R \otimes \mathbf{e_{n_2}}{}^T\mathbf{e_{n_2}},$$

where $R \otimes \mathbf{e_{n_2}}{}^T\mathbf{e_{n_2}}$ is the matrix describing the overflow of customers from queue 2 to queue 1. Unfortunately, there is no analytical solution for the generator matrix A_4.

In view of the overflow queueing system, closed form solution of the steady state probability distribution is not always available. In fact, there are a lot applications related to queueing systems whose problem size are very large [34, 35, 36, 43, 44, 52, 80]. Direct methods for solving the the probability distribution such as the Gaussian elimination and LU factorization can be found in [130, 194]. Another popular method is called the matrix analytic methods [138]. Apart from the direct methods, another class of popular numerical methods is called the iterative methods. They include those classical iterations introduced in Chapter 1 such as Jacobi method, Gauss-Seidel method and SOR method. Sometimes when the generator matrix has block structure, block Jacobi method, block Gauss-Seidel method and block SOR method are also popular methods [101]. A hybrid numerical algorithm which combines both SOR and genetic algorithm has been also introduced by Ching et al [215] for solving queueing systems. Conjugate gradient methods with circulant-based preconditioners are efficient solvers for a class of Markov chains having near-Toepltiz generator matrices. We will briefly discuss this in the following subsection.

2.1.5 The Preconditioning of Complex Queueing Systems

In many complex queueing systems, one observe both block structure, near-Toeplitz structure and sparsity in the generator matrices. Therefore iterative method such as CG method can be a good solver with a suitable preconditioner.

Circulant-based Preconditioners

In this subsection, we illustrate how to get a circulant preconditioner from a generator matrix of a queueing system. The generator matrices of the queueing networks can be written in terms of the sum of tensor products of matrices. Very often, a key block structure of a queueing system is the following: $(n + s + 1) \times (n + s + 1)$ tridiagonal matrix:

$$
Q = \begin{pmatrix}
\lambda & -\mu & & & & & & 0 \\
-\lambda & \lambda + \mu & -2\mu & & & & & \\
& \ddots & \ddots & \ddots & & & & \\
& & -\lambda & \lambda + (s-1)\mu & -s\mu & & & \\
& & & -\lambda & \lambda + s\mu & -s\mu & & \\
& & & & \ddots & \ddots & \ddots & \\
& & & & & -\lambda & \lambda + s\mu & -s\mu \\
0 & & & & & & -\lambda & s\mu
\end{pmatrix}. \qquad (2.10)
$$

This is the generator matrix of an $M/M/s/n$ queue. In this queueing system, there are s independent exponential servers, the customers arrive according to a Poisson process of rate λ and each server has a service rate of μ.

One can observe that if s is fixed and n is large then Q is close to the following tridiagonal Toeplitz matrix $\mathrm{Tri}[\lambda, -\lambda - s\mu, s\mu]$. In fact, if one considers the following circulant matrix $c(Q)$:

$$
c(Q) = \begin{pmatrix}
\lambda + s\mu & -s\mu & & & -\lambda \\
-\lambda & \lambda + s\mu & -s\mu & & \\
& \ddots & \ddots & \ddots & \\
& & -\lambda & \lambda + s\mu & -s\mu \\
-s\mu & & & -\lambda & \lambda + s\mu
\end{pmatrix}. \qquad (2.11)
$$

It is easy to see that

$$
\mathrm{rank}(c(Q) - Q) \leq s + 1
$$

independent of n for fixed s. Therefore for fixed s and large value of n, the approximate is a good one. Moreover, $c(Q)$ can be diagonalized by the discrete Fourier Transformation and closed form solution of its eigenvalues can be easily obtained. This is important in the convergence rate analysis of CG method. By applying this circulant approximation to the blocks of the generator matrices, effective preconditioners were constructed and the preconditioned systems were also proved to have singular values clustered around one, see for instance Chan and Ching [44]. A number of related applications can be found in [43, 44, 48, 50, 52, 55].

Toeplitz-Circulant-based Preconditioners

Another class of queueing systems with batch arrivals have been discussed by Chan and Ching in [43]. The generator matrices of the queueing systems of s identical exponential servers with service rate μ take the form

$$
A_n = \begin{pmatrix}
\lambda & -\mu & 0 & 0 & 0 & \cdots & 0 \\
-\lambda_1 & \lambda+\mu & -2\mu & 0 & 0 & \cdots & 0 \\
-\lambda_2 & -\lambda_1 & \lambda+2\mu & \ddots & \ddots & & \vdots \\
\vdots & -\lambda_2 & \ddots & \ddots & -s\mu & \ddots & \\
\vdots & & \ddots & \ddots & \lambda+s\mu & \ddots & 0 \\
-\lambda_{n-2} & -\lambda_{n-3} & \cdots & & \ddots & \ddots & -s\mu \\
-r_1 & -r_2 & -r_3 & \cdots & -r_{s+1} & \cdots & s\mu
\end{pmatrix}, \qquad (2.12)
$$

where r_i are such that each column sum of A_n is zero, i.e.

$$
r_i = \lambda - \sum_{k=n-i}^{\infty} \lambda_k.
$$

Here λ is the arrival rate and $\lambda_i = \lambda p_i$ where p_i is the probability that an arrived batch is of size i. It is clear that the matrix is dense and the method of circulant approximation does not work directly in this case. A Toeplitz-circulant type of preconditioner was proposed to solve this queueing system Chan and Ching [43]. The idea is that the generator matrix is close to a Toeplitz matrix whose generating function has a zero on the unit circle of order one. By factoring the zero, the quotient has no zero on the unit circle. Using this fact, a Toeplitz-circulant preconditioner is then constructed for the queueing system. Both the construction cost and the preconditioner system can be solved in $n \log(n)$ operations. Moreover, the preconditioned system was proved to have singular values clustered around one. Hence very fast convergence rate is expected when CG method is applied to solving the preconditioned system.

This idea was further applied to queueing systems with batch arrivals and negative customers Ching [54]. The term "negative customer" was first introduced by Gelenbe et al. [94, 95, 96] in the modelling of neural networks. Here the role of a negative customer is to remove a number of customers waiting in the queueing system. For example, one may consider a communication network in which messages are transmitted in a packet-switching mode. When a server fails (this corresponds to an arrival of a negative customer) during a transmission, part of the messages will be lost. One may also consider a manufacturing system where a negative customer represents a cancellation of a job. These lead to many practical applications in the modelling of physical systems.

In the queueing system, we assume that the arrival process of the batches of customers follow a Poisson process of rate λ. The batch size again follows a stationary distribution of

$$
p_i (i = 1, 2, \ldots,).
$$

Here p_i is the probability that an arrived batch is of size i. It is also assumed that the arrival process of negative customers is a Poisson process with rate

τ. The number of customers to be killed is assumed to follow a probability distribution

$$b_i (i = 1, 2, \ldots,).$$

Furthermore, if the arrived negative customer is supposed to kill i customers in the system but the number of customers in the system is less than i, then the queueing system will become empty. The killing strategy here is to remove the customers in the front of the queue, i.e. "Remove the Customers at the Head" (RCH). For $i \geq 1$, we let

$$\tau_i = b_i \tau$$

where b_i is the probability that the number of customers to be killed is i and therefore we have

$$\tau = \sum_{k=1}^{\infty} \tau_k.$$

The generator matrices of the queueing systems take the following form:

$$A_n = \begin{pmatrix}
\lambda & -u_1 & -u_2 & -u_3 & \cdots & \cdots & \cdots & -u_{n-1} \\
-\lambda_1 & \lambda + \tau + \mu & -2\mu - \tau_1 & -\tau_2 & -\tau_3 & \cdots & \cdots & -\tau_{n-2} \\
-\lambda_2 & -\lambda_1 & \lambda + \tau + 2\mu & \ddots & \ddots & \ddots & & \vdots \\
\vdots & -\lambda_2 & \ddots & \ddots & -s\mu - \tau_1 & -\tau_2 & \ddots & \vdots \\
\vdots & \vdots & \ddots & \ddots & \lambda + \tau + s\mu & \ddots & \ddots & -\tau_3 \\
\vdots & \vdots & \ddots & \ddots & \ddots & \ddots & \ddots & -\tau_2 \\
-\lambda_{n-2} & -\lambda_{n-3} & -\lambda_{n-4} & \cdots & \lambda_2 & -\lambda_1 & \lambda + \tau + s\mu & -s\mu - \tau_1 \\
-v_1 & -v_2 & -v_3 & \cdots & \cdots & -v_{n-2} & -v_{n-1} & \tau + s\mu
\end{pmatrix}.$$

Here

$$\lambda = \sum_{i=1}^{\infty} \lambda_i \quad \text{and} \quad \lambda_i = \lambda p_i$$

and

$$u_1 = \tau \quad \text{and} \quad u_i = \tau - \sum_{k=1}^{i-1} \tau_k \quad \text{for } i = 2, 3, \ldots$$

and v_i is defined such that the ith column sum is zero. The generator matrices enjoy the same near-Toeplitz structure. Toeplitz-circulant preconditioners can be constructed similarly and the preconditioned systems are proved to have singular values clustered around one, Ching [54].

Finally, we remark that there is another efficient iterative method for solving queueing systems which is not covered in the context, the multigrid methods. Interested readers may consult the following references Bramble [32], Chan et al. [45], Chang et al [47] and McCormick [163].

2.2 Search Engines

In this section, we introduce a very important algorithm used by Google in ranking the webpages in the Internet. In surfing the Internet, surfers usually use search engines to find the related webpages satisfying their queries. Unfortunately, very often there can be thousands of webpages which are relevant to the queries. Therefore a proper list of the webpages in certain order of importance is necessary. The list should also be updated regularly and frequently. Thus it is important to seek for fast algorithm for the computing the PageRank so as to reduce the time lag of updating. It turns out that this problem is difficult. The reason is not just because of the huge size of the webpages in the Internet but also the size keeps on growing rapidly.

PageRank has been proposed by Page et al. [166] to reflect the importance of each webpage, see also [223]. Larry Page and Sergey Brin are the founder of Google. In fact, one can find the following statement at Google's website [228]: "The heart of our software is PageRank^{TM}, a system for ranking web pages developed by our founders Larry Page and Sergey Brin at Stanford University. And while we have dozens of engineers working to improve every aspect of Google on a daily basis, PageRank continues to provide the basis for all of our web search tools."

A similar idea of ranking the Journals has been proposed by Garfield [98, 99] as a measure of standing for journals, which is called the *impact factor*. The impact factor of a journal is defined as the average number of citations per recently published papers in that journal. By regarding each webpage as a journal, this idea was then extended to measure the importance of the webpage in the PageRank Algorithm.

The PageRank is defined as follows. Let N be the total number of webpages in the web and we define a matrix Q called the *hyperlink matrix*. Here

$$Q_{ij} = \begin{cases} 1/k & \text{if webpage } i \text{ is an outgoing link of webpage } j; \\ 0 & \text{otherwise;} \end{cases}$$

and k is the total number of outgoing links of webpage j. For simplicity of discussion, here we assume that $Q_{ii} > 0$ for all i. This means for each webpage, there is a link pointing to itself. Hence Q can be regarded as a transition probability matrix of a Markov chain of a random walk. The analogy is that one may regard a surfer as a random walker and the webpages as the states of the Markov chain. Assuming that this underlying Markov chain is irreducible, then the steady-state probability distribution

$$(p_1, p_2, \ldots, p_N)^T$$

of the states (webpages) exists. Here p_i is the proportion of time that the random walker (surfer) visiting state (webpage) i. The higher the value of p_i is, the more important webpage i will be. Thus the PageRank of webpage i is then defined as p_i. If the Markov chain is not irreducible then one can still follow the treatment in next subsection.

An Example

We Consider a web of 3 webpages:$1, 2, 3$ such that
$1 \rightarrow 1,\ 1 \rightarrow 2,\ 1 \rightarrow 3$
$2 \rightarrow 1,\ 2 \rightarrow 2,$
$3 \rightarrow 2,\ 3 \rightarrow 3.$

One can represent the relationship by the following Markov chain.

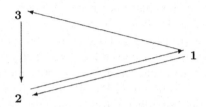

Fig. 2.4. An example of three webpages.

The transition probability matrix of this Markov chain is then given by

$$Q = \begin{array}{c} 1 \\ 2 \\ 3 \end{array} \begin{pmatrix} 1/3 & 1/2 & 0 \\ 1/3 & 1/2 & 1/2 \\ 1/3 & 0 & 1/2 \end{pmatrix}.$$

The steady state probability distribution of the Markov chain

$$\mathbf{p} = (p_1, p_2, p_3)$$

satisfies

$$\mathbf{p} = Q\mathbf{p} \quad \text{and} \quad p_1 + p_2 + p_3 = 1.$$

Solving the above linear system, we get

$$(p_1, p_2, p_3) = (\frac{3}{9}, \frac{4}{9}, \frac{2}{9}).$$

Therefore the ranking of the webpages is:

Webpage 2 > Wepbage 1 > Webpage 3.

One can also interpret the result as follows. Both 1 and 3 point to 2 and therefore 2 is the most important. Since 2 points to 1 but not 3, 1 is more important then 3.

Since the size of the Markov chain is huge and the time for computing the PageRank required by Google is just a few days, direct method for solving the steady-state probability is not desirable. Iterative methods Baldi et al. [12] and decomposition methods Avrachenkov and Litvak [9] have been proposed to solve the problem. Another pressing issue is that the size of the webpages grows rapidly, and the PageRank of each webpage has to be updated regularly. Here we seek for adaptive and parallelizable numerical algorithms for solving the PageRank problem. One potential method is the hybrid iterative method proposed in Yuen et al. [215]. The hybrid iterative method was first proposed by He et al. [107] for solving the numerical solutions of PDEs and it has been also successfully applied to solving the steady-state probability distributions of queueing networks [215]. The hybrid iterative method combines the evolutionary algorithm and the Successive Over-Relaxation (SOR) method. The evolutionary algorithm allows the relaxation parameter w to be adaptive in the SOR method. Since the cost of SOR method per iteration is more expansive and less efficient in parallel computing for our problem (as the matrix system is huge), here we will also consider replacing the role of SOR method by the Jacobi Over-Relaxation (JOR) method [101, 130]. The reason is that JOR method is easier to be implemented in parallel computing environment. Here we present hybrid iterative methods based on SOR/JOR and evolutionary algorithm. The hybrid method allows the relaxation parameter w to be adaptive in the SOR/JOR method. We give a brief mathematical discussion on the PageRank approach. We then briefly describe the power method, a popular approach for solving the PageRank.

2.2.1 The PageRank Algorithm

The PageRank Algorithm has been used successfully in ranking the importance of web-pages by Google [223]. Consider a web of N webpages with Q being the hyperlink matrix. Since the matrix Q can be reducible, to tackle this problem, one can consider the revised matrix P:

$$P = \alpha \begin{pmatrix} Q_{11} & Q_{12} & \cdots & Q_{1N} \\ Q_{21} & Q_{22} & \cdots & Q_{2N} \\ \vdots & \vdots & \vdots & \vdots \\ Q_{N1} & Q_{N2} & \cdots & Q_{NN} \end{pmatrix} + \frac{(1-\alpha)}{N} \begin{pmatrix} 1 & 1 & \cdots & 1 \\ 1 & 1 & \cdots & 1 \\ \vdots & \vdots & \vdots & \vdots \\ 1 & 1 & \cdots & 1 \end{pmatrix} \tag{2.13}$$

where $0 < \alpha < 1$. In this case, the matrix P is irreducible and aperiodic, therefore the steady state probability distribution exists and is unique [180]. Typical values for α are 0.85 and $(1-1/N)$, see for instance [12, 223, 106]. The value $\alpha = 0.85$ is a popular one because power method works very well for this problem [106]. However, this value can be considered to be too small and may distort the original ranking of the webpages, see the example in Section 2.2.3.

One can interpret (2.13) as follows. The idea of the algorithm is that, for a network of N webpages, each webpage has an inherent importance of $(1 - \alpha)/N$. If a page P_i has an importance of p_i, then it will contribute an importance of αp_i which is shared among the webpages that it points to. The importance of webpage P_i can be obtained by solving the following linear system of equations subject to the normalization constraint:

$$
\begin{pmatrix} p_1 \\ p_2 \\ \vdots \\ p_N \end{pmatrix} = \alpha \begin{pmatrix} Q_{11} & Q_{12} & \cdots & Q_{1N} \\ Q_{21} & Q_{22} & \cdots & Q_{2N} \\ \vdots & \vdots & \vdots & \vdots \\ Q_{N1} & Q_{N2} & \cdots & Q_{NN} \end{pmatrix} \begin{pmatrix} p_1 \\ p_2 \\ \vdots \\ p_N \end{pmatrix} + \frac{(1-\alpha)}{N} \begin{pmatrix} 1 \\ 1 \\ \vdots \\ 1 \end{pmatrix}. \qquad (2.14)
$$

Since

$$
\sum_{i=1}^{N} p_i = 1,
$$

(2.14) can be re-written as

$$
(p_1, p_2, \ldots, p_N)^T = P(p_1, p_2, \ldots, p_N)^T.
$$

2.2.2 The Power Method

The power method is a popular method for solving the PageRank problem. The power method is an iterative method for solving the largest eigenvalue in modulus (the dominant eigenvalue) and its corresponding eigenvector [101]. The idea of the power method can be briefly explained as follows. Given an $n \times n$ matrix A and suppose that (i) there is a single eigenvalue of maximum modulus and the eigenvalues $\lambda_1, \lambda_2, \cdots, \lambda_n$ be labelled such that

$$
|\lambda_1| > |\lambda_2| \geq |\lambda_3| \geq \cdots \geq |\lambda_n|;
$$

(ii) there is a linearly independent set of n unit eigenvectors. This means that there is a basis

$$
\left\{ \mathbf{u}^{(1)}, \mathbf{u}^{(2)}, \ldots, \mathbf{u}^{(n)} \right\}
$$

such that

$$
A\mathbf{u}^{(i)} = \lambda_i \mathbf{u}^{(i)}, \quad i = 1, 2, \ldots, n, \quad \text{and} \quad \|\mathbf{u}^{(i)}\| = 1.
$$

Then begin with an initial vector $\mathbf{x}^{(0)}$, one may write

$$
\mathbf{x}^{(0)} = a_1 \mathbf{u}^{(1)} + a_2 \mathbf{u}^{(2)} + \cdots + a_n \mathbf{u}^{(n)}.
$$

Now we iterate the initial vector with the matrix A as follows:

$$
A^k \mathbf{x}^{(0)} = a_1 A^k \mathbf{u}^{(1)} + \ldots + a_n A^k \mathbf{u}^{(n)} = a_1 \lambda_1^k \mathbf{u}^{(1)} + \ldots + a_n \lambda_n^k \mathbf{u}^{(n)}
$$

$$
= \lambda_1^k \left\{ a_1 \mathbf{u}^{(1)} + \left(\frac{\lambda_2}{\lambda_1} \right)^k a_n \mathbf{u}^{(2)} + \ldots + \left(\frac{\lambda_n}{\lambda_1} \right)^k a_n \mathbf{u}^{(n)} \right\}.
$$

Since

$$\frac{|\lambda_i|}{|\lambda_1|} < 1 \quad \text{for } i = 2, \ldots, n,$$

we have

$$\lim_{k \to \infty} \frac{|\lambda_i|^k}{|\lambda_1|^k} = 0 \quad \text{for } i = 2, \ldots, n.$$

Hence we have

$$A^k \mathbf{x}^{(0)} \approx a_1 \lambda_1^k \mathbf{u}^{(1)}.$$

To get an approximation for $\mathbf{u}^{(1)}$ we introduce a normalization in the iteration:

$$\mathbf{r}_{k+1} = \frac{A^{k+1}\mathbf{x}^{(0)}}{\|A^k \mathbf{x}^{(0)}\|_2}$$

then we have

$$\lim_{k \to \infty} \mathbf{r}_{k+1} = \lim_{k \to \infty} \frac{a_1 \lambda_1^{k+1} \mathbf{u}^{(1)}}{\|a_1 \lambda_1^k \mathbf{u}^{(1)}\|_2} = \lambda_1 \mathbf{u}^{(1)}.$$

It turns out that for the PageRank problem, the largest eigenvalue of P is 1 and the corresponding eigenvector in normalized form is the PageRank vector. The main computational cost of this method comes from the matrix-vector multiplications. The convergence rate of the power method depends on the ratio of $|\lambda_2/\lambda_1|$ where λ_1 and λ_2 are respectively the largest and the second largest eigenvales of the matrix P. It was proved by Haveliwala and Kamvar [106] that for the second largest eigenvalue of P, we have

$$|\lambda_2| \le \alpha \quad \text{for} \quad 0 \le \alpha \le 1.$$

Since $\lambda_1 = 1$, the convergence rate of the power method is α, see for instance [101]. A popular value for α is 0.85. With this value, it was mentioned in Kamvar et al. [123] that the power method on a web data set of over 80 million pages converges in about 50 iterations.

2.2.3 An Example

In this subsection, we consider a small example of six webpages. This example demonstrates that the value of $\alpha = 0.85$ can be too small and distort the true ranking of the webpages even if the web size is small. In the example, the webpages are organized as follows:

Webpage 1 $\rightarrow$ 1, 3, 4, 5.
Webpage 2 $\rightarrow$ 2, 3, 5, 6.
Webpage 3 $\rightarrow$ 1, 2, 3, 4, 5, 6.
Webpage 4 $\rightarrow$ 2, 3, 4, 5.
Webpage 5 $\rightarrow$ 1, 3, 5.
Webpage 6 $\rightarrow$ 1, 6.

From the given structure of the webpages, we have the hyperlink matrix as follows:

$$Q = \begin{pmatrix} 0.2500 & 0.0000 & 0.1667 & 0.0000 & 0.3333 & 0.5000 \\ 0.0000 & 0.2500 & 0.1667 & 0.2500 & 0.0000 & 0.0000 \\ 0.2500 & 0.2500 & 0.1667 & 0.2500 & 0.3333 & 0.0000 \\ 0.2500 & 0.0000 & 0.1667 & 0.2500 & 0.0000 & 0.0000 \\ 0.2500 & 0.2500 & 0.1667 & 0.2500 & 0.3333 & 0.0000 \\ 0.0000 & 0.2500 & 0.1667 & 0.0000 & 0.0000 & 0.5000 \end{pmatrix}$$

then the steady state probability distribution is given by

$$(0.2260, 0.0904, 0.2203, 0.1243, 0.2203, 0.1186)^T$$

and the ranking should be $1 > 3 \geq 5 > \underline{4 > 6} > 2$. For $\alpha = 0.85$, we have

$$P = \begin{pmatrix} 0.2375 & 0.0250 & 0.1667 & 0.0250 & 0.3083 & 0.4500 \\ 0.0250 & 0.2375 & 0.1667 & 0.2375 & 0.0250 & 0.0250 \\ 0.2375 & 0.2375 & 0.1667 & 0.2375 & 0.3083 & 0.0250 \\ 0.2375 & 0.0250 & 0.1667 & 0.2375 & 0.0250 & 0.0250 \\ 0.2375 & 0.2375 & 0.1667 & 0.2375 & 0.3083 & 0.0250 \\ 0.0250 & 0.2375 & 0.1667 & 0.0250 & 0.0250 & 0.4500 \end{pmatrix}.$$

In this case, the steady state probability distribution is given by

$$(0.2166, 0.1039, 0.2092, 0.1278, 0.2092, 0.1334)^T$$

and the ranking should be $1 > 3 \geq 5 > \underline{6 > 4} > 2$. We observe that the ranking of states 6 and 4 are inter-changed in the two approaches.

2.2.4 The SOR/JOR Method and the Hybrid Method

In this section, we present a hybrid algorithm for solving the steady state probability of a Markov chain, Yuen et al. [215, 216]. We first give a review on the JOR method for solving linear system, in particular solving the steady state probability distribution of a finite Markov chain. We then introduce the hybrid algorithm based on the SOR/JOR method and the evolutionary algorithm. For the SOR method, it has been discussed in Chapter one. Now we consider a non-singular linear system $B\mathbf{x} = \mathbf{b}$, the JOR method is a classical iterative method. The idea of JOR method can be explained as follows. We write $B = D - (D - B)$ where D is the diagonal part of the matrix B. Given an initial guess of the solution, $\mathbf{x}_0$, the JOR iteration scheme reads:

$$\begin{aligned} \mathbf{x}_{n+1} &= (I - wD^{-1}B)\mathbf{x}_n + wD^{-1}\mathbf{b} \\ &\equiv B_w\mathbf{x}_n + wD^{-1}\mathbf{b}. \end{aligned} \tag{2.15}$$

The parameter w is called the relaxation parameter and it lies between 0 and 1 [11]. Clearly if the scheme converges, the limit will be the solution of

$$B\mathbf{x} = \mathbf{b}.$$

The choice of the relaxation parameter w affects the convergence rate of the SOR/JOR method very much, see for instance [215, 216]. In general, the optimal value of w is unknown. For more details about the SOR/JOR method and its property, we refer readers to [11, 101].

The generator matrix P of an irreducible Markov chain is singular and has a null space of dimension one (the null vector corresponds to the steady state probability distribution). One possible way to solve the steady state probability distribution is to consider the following revised system:

$$A\mathbf{x} = (P + \mathbf{e}_n^T \mathbf{e}_n)\mathbf{x} = \mathbf{e}_n^T \tag{2.16}$$

where $\mathbf{e}_n = (0, 0, \ldots, 0, 1)$ is a unit vector. The steady state probability distribution is then obtained by normalizing the solution $\mathbf{x}$, see for instance Ching [52]. We remark that the linear system (2.16) is irreducibly diagonal dominant. The hybrid method based on He et al. [107] and Yuen et al. [215] consists of four major steps: *initialization, mutation, evaluation and adaptation.*

In the initialization step, we define the size of the population k of the approximate steady-state probability distribution. This means that we also define k approximates to initialize the algorithm. Then use the JOR iteration in (2.15) as the "mutation step". In the evaluation step, we evaluate how "good" each member in the population is by measuring their residuals. In this case, it is clear that the smaller the residual the better the approximate and therefore the better the member in the population. In the adaptation step, the relaxation parameters of the "weak" members are migrated (with certain probability) towards the best relaxation parameter. The hybrid algorithm reads:

Step 1: Initialization: We first generate an initial population of k ($2 \leq k \leq n$) identical steady-state probability distributions as follows:

$$\{\mathbf{e}_i : i = 1, 2, \ldots, k\}$$

where $\mathbf{e}_i = (1, 1, \ldots, 1)$. We then compute

$$r_i = ||B\mathbf{e}_i - \mathbf{b}||_2$$

and define a set of relaxation parameters $\{w_1, w_2, \ldots, w_k\}$ such that

$$w_i = \tau + \frac{(1 - 2\tau)(k - i)}{k - 1}, \quad i = 1, 2, \ldots, k.$$

Here $\tau \in (0, 1)$ and therefore $w_i \in [\tau, 1 - \tau]$. We set $\tau = 0.01$ in our numerical experiments. We then obtain a set of ordered triples

$$\{(\mathbf{e}_i, w_i, r_i) : i = 1, 2, \ldots, k\}.$$

Step 2: Mutation: The mutation step is carried out by doing a SOR/JOR iteration on each member $\mathbf{x}_i$ ($\mathbf{x}_i$ is used as the initial in the SOR/JOR) of the population with their corresponding w_i. We then get a new set of approximate steady-state probability distributions: $\mathbf{x}_i$ for $i = 1, 2, \ldots, k$. Hence we have a new set of

$$\{(\mathbf{x}_i, w_i, r_i) : i = 1, 2, \ldots, k\}.$$

Goto Step 3.

Step 3: Evaluation: For each $\mathbf{x}_i$, we compute and update its residual

$$r_i = ||B\mathbf{x}_i - \mathbf{b}||_2.$$

This is used to measure how "good" an approximate $\mathbf{x}_i$ is. If $r_j < tol$ for some j then stop and output the approximate steady state probability distribution $\mathbf{x}_j$. Otherwise we update r_i of the ordered triples

$$\{(\mathbf{x}_i, w_i, r_i) : i = 1, 2, \ldots, k\}$$

and goto Step 4.

Step 4: Adaptation: In this step, the relaxation factors w_k of the weak members (relatively large r_i) in the population are moving towards the best one with certain probability. This process is carried out by first performing a linear search on $\{r_i\}$ to find the best relaxation factor, w_j. We then adjust all the other w_k as follows:

$$w_k = \begin{cases} (0.5 + \delta_1) * (w_k + w_j) & \text{if } (0.5 + \delta_1) * (w_k + w_j) \in [\tau, 1 - \tau] \\ w_k & \text{otherwise,} \end{cases}$$

where δ_1 is a random number in $[-0.01, 0.01]$. Finally the best w_j is also adjusted by

$$w_j = \delta_2 * w_j + (1 - \delta_2) * \frac{(w_1 + w_2 + \ldots + w_{j-1} + w_{j+1} + \ldots + w_k)}{k - 1}$$

where δ_2 is a random number in $[0.99, 1]$. A new set of $\{w_i\}$ is then obtained and hence

$$\{(\mathbf{x}_i, w_i, r_i) : i = 1, 2, \ldots, k\}.$$

Goto Step 2.

2.2.5 Convergence Analysis

In this section, we consider the linear system $B\mathbf{x} = \mathbf{b}$ where B is strictly diagonal dominant, i.e.

$$|B_{ii}| > \sum_{j=1, j\neq i}^{N} |B_{ij}| \quad \text{for} \quad i = 1, 2, \ldots, N$$

where N is the size of the matrix.

We first prove that the hybrid algorithm with SOR method converges for a range of w. We begin with the following lemma.

Lemma 2.3. *Let B be a strictly diagonal dominant square matrix and*

$$K = \max_i \left\{ \sum_{j=1, j\neq i}^{m} \frac{|B_{ij}|}{|B_{ii}|} \right\} < 1,$$

then

$$||B_w||_\infty < 1 \quad \text{for} \quad 0 < w < 2/(1+K)$$

where B_w is defined in (2.13).

Proof. Let $\mathbf{x}$ be an $n \times 1$ vector such that $||\mathbf{x}||_\infty = 1$. We are going to prove that

$$||B_w\mathbf{x}||_\infty \leq 1 \quad \text{for} \quad 0 < w < 2/(1+K).$$

Consider

$$\mathbf{y} = (D - wL)^{-1}((1-w)D + wU)\mathbf{x}$$

and we have

$$(D - wL)\mathbf{y} = ((1-w)D + wU)\mathbf{x}$$

i.e.,

$$\begin{pmatrix} B_{11} & 0 & \cdots & & \cdots & 0 \\ -wB_{21} & B_{22} & \ddots & & & \vdots \\ \vdots & & \ddots & \ddots & & \vdots \\ \vdots & & & \ddots & & 0 \\ -wB_{m1} & \cdots & \cdots & -wB_{m,m-1} & B_{mm} \end{pmatrix} \begin{pmatrix} y_1 \\ y_2 \\ \vdots \\ \vdots \\ y_m \end{pmatrix}$$

$$= \begin{pmatrix} (1-w)B_{11} & wB_{12} & \cdots\cdots & & wB_{1m} \\ 0 & (1-w)B_{22} & \ddots & & \vdots \\ \vdots & & \ddots & \ddots & \vdots \\ \vdots & & & \ddots & wB_{m-1,m} \\ 0 & \cdots & \cdots & 0 & (1-w)B_{mm} \end{pmatrix} \begin{pmatrix} x_1 \\ x_2 \\ \vdots \\ \vdots \\ x_m \end{pmatrix}.$$

Case 1: $1 \leq w < 2/(K+1)$.

For the first equation, we have

$$B_{11}y_1 = (1-w)B_{11}x_1 + w\sum_{j=2}^{m}B_{1j}x_j.$$

Since

$$|x_i| \le 1 \quad \text{and} \quad \sum_{j=2}^{m}|B_{1j}| < K|B_{11}|,$$

we have

$$|y_1| \le |1-w| + wK = w(1+K) - 1 < 1.$$

For the second equation, we have

$$B_{22}y_2 = (1-w)B_{22}x_2 + wB_{21}y_1 + w\sum_{j=3}^{m}B_{2j}x_j.$$

Since

$$|y_1| \le 1,\ |x_i| \le 1 \quad \text{and} \quad \sum_{j=1,j\neq 2}^{m}|B_{2j}| < K|B_{22}|,$$

we have

$$|y_2| \le |1-w| + wK = w(1+K) - 1 < 1.$$

Inductively, we have $|y_i| < 1$ and hence $\|\mathbf{y}\|_\infty < 1$. Therefore we proved that

$$\|B_w\|_\infty < 1 \quad \text{for} \quad 1 \le w < 2/(1+K).$$

Case 2: $0 < w < 1$.

For the first equation, we have

$$B_{11}y_1 = (1-w)B_{11}x_1 + w\sum_{j=2}^{m}B_{1j}x_j.$$

Since

$$|x_i| \le 1 \quad \text{and} \quad \sum_{j=2}^{m}|B_{1j}| < |B_{11}|,$$

we have

$$|y_1| < 1 - w + w = 1.$$

For the second equation, we have

$$B_{22}y_2 = (1-w)B_{22}x_2 + wB_{21}y_1 + w\sum_{j=3}^{m}B_{2j}x_j.$$

Since

$$|y_1| \le 1,\ |x_i| \le 1 \quad \text{and} \quad \sum_{j=1,j\neq 2}^{m}|B_{2j}| < |B_{22}|,$$

we have
$$|y_2| < 1 - w + w = 1.$$
Inductively, we have $|y_i| < 1$ and hence $||\mathbf{y}||_\infty < 1$. Therefore
$$||B_w||_\infty < 1 \quad \text{for} \quad 0 < w < 1.$$
Combining the results, we have
$$||B_w||_\infty < 1 \quad \text{for} \quad 0 < w < 2/(1 + K).$$

Proposition 2.4. *The hybrid algorithm converges for* $w \in [\tau, 2/(1+K) - \tau]$ *where* $0 < \tau < 1/(1+K)$.

Proof. We note that
$$f(\tau) = \max_{w \in [\tau, 2/(1+K) - \tau]} \{||(B_w)||_\infty\}$$
exists and less than one and let us denote it by $0 \le f(\tau) < 1$. Therefore in each iteration of the hybrid method, the matrix norm ($||.||_\infty$) of the residual is decreased by a fraction not less than $f(\tau)$. By using the fact that
$$||ST||_\infty \le ||S||_\infty ||T||_\infty,$$
the hybrid algorithm is convergent.

We then prove that the hybrid algorithm with JOR method converges for a range of w. We have the following lemma.

Lemma 2.5. *Let* B *be a strictly diagonal dominant square matrix and*
$$K = \max_i \left\{ \sum_{j=1, j \ne i}^{N} \frac{|B_{ji}|}{|B_{ii}|} \right\} < 1,$$
then
$$||B_w||_1 \le 1 - (1 - K)w < 1 \quad \text{for} \quad \tau < w < 1 - \tau$$
where B_w *is defined in (2.13).*

By using the similar approach in as in Proposition 2.4, one can prove that

Proposition 2.6. *The hybrid iterative method converges for* $w \in [\tau, 1 - \tau]$.

Proof. We observe that
$$f(\tau) = \max_{w \in [\tau, 1 - \tau]} \{||B_w||_1\}$$
exists and less than one and let us denote it by $0 \le f(\tau) < 1$. Therefore in each iteration of the hybrid method, the matrix norm ($||.||_1$) of the residual is decreased by a fraction not less than $f(\tau)$. By using the fact that
$$||ST||_1 \le ||S||_1 ||T||_1,$$
the hybrid algorithm is convergent.

We note that the matrix A in (2.14) is irreducibly diagonal dominant only but not strictly diagonal dominant. Therefore the condition in Lemma 2.3 and 2.5 is not satisfied. However, one can always consider a regularized linear system as follows:

$$(A + \epsilon I)\mathbf{x} = \mathbf{b}.$$

Here I is the identity matrix and $\epsilon > 0$ can be chosen as small as possible. Then the matrix $(A + \epsilon I)$ is strictly diagonal dominant but this will introduce a small error of $O(\epsilon)$ to the linear system. Numerical results in Yuen et al. [215, 216] indicate that the hybrid algorithm is very efficient in solving steady state distribution of queueing systems and ranking webpages in the Web. Here we present some small scale numerical results (three different data sets) for two typical values of α in Tables 2.1 and 2.2 (Taken from [216]). Here k is the size of population and N is the number of webpages.

Table 2.1. Number of iterations for convergence ($\alpha = 1 - 1/N$).

JOR	Data Set	1		Data Set	2		Data Set	3		
N	100	200	300	400	100	200	300	400	100	200	300	400
$k = 2$	41	56	42	42	57	95	58	70	31	26	32	25
$k = 3$	56	60	42	42	56	75	57	61	31	35	43	25
$k = 4$	46	59	42	42	55	72	58	62	31	32	38	25
$k = 5$	56	60	43	43	56	68	57	60	32	30	36	26

SOR	Data Set	1		Data Set	2		Data Set	3		
N	100	200	300	400	100	200	300	400	100	200	300	400
$k = 2$	20	18	17	17	16	15	16	15	18	14	19	15
$k = 3$	30	27	17	25	16	23	16	23	18	21	29	15
$k = 4$	25	24	19	22	17	21	16	21	18	19	26	18
$k = 5$	30	28	19	23	17	21	16	20	20	20	25	17

2.3 Summary

In this chapter, we discussed two important applications of Markov chain, the classical Markovian queueing networks and the Modern PageRank algorithm. For the latter application, in fact, it comes from the measurement of prestige in a network. The computation of prestige in a network is an important issue Bonacich and Lloyd [25, 26] and it has many other applications such as social networks Wasserman and Faust [206] and disease transmission, Bell et al. [15]. A number of methods based on the computation of eigenvectors have been proposed in the literatures, see for instance Langville and Meyer [137]. Further research can be done in developing models and algorithms for the case when

Table 2.2. Number of iterations for convergence ($\alpha = 0.85$).

JOR	Data Set		1		Data Set		2		Data Set		3
N	**100**	**200**	**300**	**400**	**100**	**200**	**300**	**400**	**100**	**200**	**300**	**400**
$k = 2$	42	56	44	47	61	82	66	64	18	28	32	26
$k = 3$	55	60	45	52	62	81	63	62	18	36	42	26
$k = 4$	53	59	45	49	58	71	62	62	18	33	38	26
$k = 5$	53	65	45	49	61	70	64	62	18	32	37	26

SOR	Data Set		1		Data Set		2		Data Set		3
N	**100**	**200**	**300**	**400**	**100**	**200**	**300**	**400**	**100**	**200**	**300**	**400**
$k = 2$	19	17	17	16	16	14	15	15	15	14	19	16
$k = 3$	28	26	17	24	16	22	15	23	15	23	29	16
$k = 4$	24	23	19	21	16	20	16	21	17	20	25	16
$k = 5$	28	26	19	21	17	21	16	20	16	20	23	16

there are negative relations in the network, Tai et al. [195]. In a network, being chosen or nominated by a popular or powerful person (webpage) would add one's popularity. Instead of supporting a member, a negative relation means being against by a member in the network.

3

Re-manufacturing Systems

3.1 Introduction

In this chapter, the inventory controls of demands and returns of single-item inventory systems is discussed. In fact, there are many research papers on inventory control of repairable items and returns, most of them describe the system as a closed-loop queueing network with constant number of items inside [78, 158, 201]. Disposal of returns [127, 200] is allowed in the models presented here. The justification for disposal is that accepting all returns will lead to extremely high inventory level and hence very high inventory cost. Sometimes transshipment of returns is allowed among the inventory systems to reduce the rejection rate of returns. Other re-manufacturing models can be found in [117, 200, 196] and good reviews and current advances of the related topics can be found in [23, 84, 92, 132, 157].

As a modern marketing strategy to encourage the customers to buy products, the customers are allowed to return the bought product with full refund within a period of one week. As a result, many customers may take advantage of this policy and the manufacturers have to handle a lot of such returns. Very often, the returns are still in good condition, and can be put back to the market after checking and packaging. The first model we introduce here attempt to model this situation. The model is a single-item inventory system for handling returns is captured by using a queueing network. In this model, the demands and the returns are assumed to follow two independent Poisson processes. The returns are tested and repaired with the standard requirements. Repaired returns will be put into the serviceable inventory and non-repairable returns will be disposed. The repairing time is assumed to be negligible. A similar inventory model with returns has been discussed in [110]. However, the model in [110] includes neither the replenishment costs nor the transshipment of returns. In this model, the inventory system is controlled by a popular (r, Q) continuous review policy. The inventory level of the serviceable product is modelled as an irreducible continuous time Markov chain.

The generator matrix for the model is given and a closed form solution for the system steady state probability distribution is also derived.

Next, two independent identical inventory systems are considered and transshipment of returns from one inventory system to another is allowed. The joint inventory levels of the serviceable product is modelled as a two-dimensional irreducible continuous time Markov chain. The generator matrix for this advanced model is given and a closed form approximation of the solution of the system steady state probability distribution is derived. Analysis of the average running cost of the joint inventory system can be carried out by using the approximated probability distribution. The focus is on the inventory cost and the replenishment cost of the system because the replenishment lead time is assumed to be zero and there is no backlog or loss of demands. It is shown that in the transshipment model, the rejection rate of the returns is extremely small and decreases significantly when the re-order size $(Q + 1)$ is large. The model is then extended to multiple inventory/return systems with a single depot. This kind of model is of particular interest when the re-manufacturer has several re-cycling locations. Since the locations can be easily connected by an information network, excessive returns can be forwarded to the nearby locations or to the main depot directly. This will greatly cut down the disposal rate. The handling of used machines in IBM (a big recovery network) serves as a good example for the application of this model [92]. More examples and related models can be found in [92, pp. 106-131].

Finally, a hybrid system consists of a re-manufacturing process and a manufacturing process is discussed. The hybrid system captures the re-manufacturing process and the system can produce serviceable product when the return rate is zero.

The remainder of this chapter is organized as follows. In Section 3.2, a single-item inventory model for handling returns is presented. In Section 3.3, the model is extended to the case that lateral transshipment of returns is allowed among the inventory systems. In Section 3.4, we discuss a hybrid re-manufacturing system. Finally, concluding remarks are given in Section 3.5.

3.2 An Inventory Model for Returns

In this section, a single-item inventory system is presented. The demands and returns of the product are assumed to follow two independent Poisson processes with mean rates λ and μ respectively. The maximum inventory capacity of the system is Q. When the inventory level is Q, any arrived return will be disposed. A returned product is checked/repaired before putting into the serviceable inventory. Here it is assumed that only a stationary proportion, let us say $a \times 100\%$ of the returned product is repairable and a non-repairable return will be disposed. The checking/repairing time of a returned product is assumed to be negligible. The notations for later discussions is as follows:

(i) λ^{-1}, the mean inter-arrival time of demands,

(ii) μ^{-1}, the mean inter-arrival time of returns,
(iii) a, the probability that a returned product is repairable,
(iv) Q, maximum inventory capacity,
(v) I, unit inventory cost,
(vi) R, cost per replenishment order.

An (r, Q) inventory control policy is employed as inventory control. Here, the lead time of a replenishment is assumed to be negligible. For simplicity of discussion, here we assume that $r = 0$. In a traditional $(0, Q)$ inventory control policy, a replenishment size of Q is placed whenever the inventory level is 0. Here, we assume that there is no loss of demand in our model. A replenishment order of size $(Q+1)$ is placed when the inventory level is 0 and there is an arrived demand. This will then clear the arrived demand and bring the inventory level up to Q, see Fig. 3.1 (Taken from [76]). In fact, State '−1' does not exist in the Markov chain, see Fig. 3.2 (Taken from [76]) for instance.

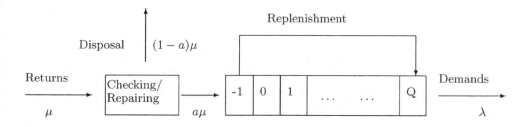

Fig. 3.1. The single-item inventory model.

The states of the Markov chain are ordered according to the inventory levels in ascending order and get the following Markov chain.

The $(Q+1) \times (Q+1)$ system generator matrix is given as follows:

$$A = \begin{matrix} 0 \\ 1 \\ \vdots \\ \vdots \\ Q \end{matrix} \begin{pmatrix} \lambda + a\mu & -\lambda & & & 0 \\ -a\mu & \lambda + a\mu & -\lambda & & \\ & \ddots & \ddots & \ddots & \\ & & -a\mu & \lambda + a\mu & -\lambda \\ -\lambda & & & -a\mu & \lambda \end{pmatrix}. \qquad (3.1)$$

The steady state probability distribution $\mathbf{p}$ of the system satisfies

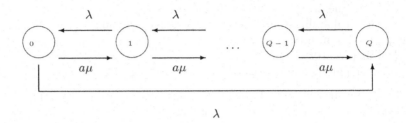

Fig. 3.2. The Markov chain.

$$\mathbf{Ap} = \mathbf{0} \quad \text{and} \quad \mathbf{1}^T \mathbf{p} = 1. \tag{3.2}$$

By direct verification the following propositions and corollary were obtained.

Proposition 3.1. *The steady state probability distribution* $\mathbf{p}$ *is given by*

$$p_i = K(1 - \rho^{i+1}), i = 0, 1, \ldots, Q \tag{3.3}$$

where

$$\rho = \frac{a\mu}{\lambda} \quad \text{and} \quad K = \frac{1 - \rho}{(1 + Q)(1 - \rho) - \rho(1 - \rho^{Q+1})}.$$

By using the result of the steady state probability in Proposition 3.1, the following corollary is obtained.

Corollary 3.2. *The expected inventory level is*

$$\sum_{i=1}^{Q} i p_i = \sum_{i=1}^{Q} K(i - i\rho^{i+1}) = K\left(\frac{Q(Q+1)}{2} + \frac{Q\rho^{Q+2}}{1-\rho} - \frac{\rho^2(1-\rho^Q)}{(1-\rho)^2}\right),$$

the average rejection rate of returns is

$$\mu p_Q = \mu K(1 - \rho^{Q+1})$$

and the mean replenishment rate is

$$\lambda \times p_0 \times \frac{\lambda^{-1}}{\lambda^{-1} + (a\mu)^{-1}} = \frac{\lambda K(1-\rho)\rho}{(1+\rho)}.$$

Proposition 3.3. *If* $\rho < 1$ *and* Q *is large then*

$$K \approx (1 + Q)^{-1}$$

and the approximated average running cost (inventory and replenishment cost) is

$$C(Q) \approx \frac{QI}{2} + \frac{\lambda(1-\rho)\rho R}{(1+\rho)(1+Q)}.$$

The optimal replenishment size is

$$Q^* + 1 \approx \sqrt{\frac{2\lambda(1-\rho)\rho R}{(1+\rho)I}} = \sqrt{\frac{2a\mu R}{I}\left(\frac{2\lambda}{\lambda+a\mu}-1\right)}. \tag{3.4}$$

One can observe that the optimal replenishment size Q^* increases if λ, R increases or I decreases.

We end this section by the following remarks.

- The model can be extended to multi-item case when there is no limit in the inventory capacity. The trick is to use independent queueing networks to model individual products. Suppose there are s different products and their demand rates, return rates, unit inventory costs, cost per replenishment order and the probability of getting a repairable return are given by $\lambda_i, \mu_i, I_i, R_i$ and a_i respectively. Then the optimal replenishment size of each product i will be given by (3.4)

$$Q_i^* + 1 \approx \sqrt{\frac{2a_i\mu_i R_i}{I_i}\left(\frac{2\lambda_i}{\lambda_i+a_i\mu_i}-1\right)} \qquad \text{for} \quad i = 1, 2, \ldots, s.$$

- To include the inventory capacity in the system. In this case, one can have approximations for the steady state probability distributions for the inventory levels of the returns and the serviceable product if it is assumed that capacity for storing returns is large. Then the inventory levels of the returns form an M/M/1 queue and the output process of an M/M/1 queue in steady-state is again a Poisson process with same mean rate, see the lemma below.

Lemma 3.4. *The output process of an M/M/1 queue in steady state is again a Poisson process with same mean as the input rate.*

Proof. We first note that if X and Y be two independent exponential random variables with means λ^{-1} and μ^{-1} respectively. Then the probability density function for the random variable $Z = X + Y$ is given by

$$f(z) = \frac{\lambda\mu}{\mu-\lambda}e^{-\lambda z} - \frac{\lambda\mu}{\mu-\lambda}e^{-\mu z}.$$

Let the arrival rate of the M/M/1 queue be λ and the service rate of the server be μ. There are two cases to be considered: the server is idle (the steady-state probability is $(1 - \lambda/\mu)$ by (see Chapter 2) and the server is not idle (the steady state probability is λ/μ.)

For the former case, the departure time follows $f(z)$ (a waiting time for an arrival plus a service time). For the latter case, the departure time follows

$\mu e^{-\mu z}$. Thus the probability density function $g(z)$ for the departure time is given by

$$(1 - \frac{\lambda}{\mu})f(z) + \frac{\lambda}{\mu}(\mu e^{-\mu z}) = \frac{\lambda\mu}{\mu - \lambda}e^{-\lambda z} - \frac{\lambda\mu}{\mu - \lambda}e^{-\mu z}$$
$$- \frac{\lambda^2}{\mu - \lambda}e^{-\lambda z} + \frac{\lambda^2}{\mu - \lambda}e^{-\mu z} + \lambda e^{-\mu z}.$$

Thus

$$g(z) = \lambda e^{-\lambda z}$$

is the exponential distribution. This implies that the departure process is a Poisson process. Because from Proposition 1.35, the departure process is a Poisson process with mean λ if and only if the inter-departure time follows the exponential distribution with mean λ^{-1}.

- One can also take into account the lead time of a replenishment and the checking/repairing time of a return. In this case, it becomes a tandem queueing network and the analytic solution for the system steady state probability distribution is not available in general. Numerical method based on preconditioned conjugate gradient method has been applied to solve this type of tandem queueing system, see for instance [43, 44, 48, 50, 52, 55].

3.3 The Lateral Transshipment Model

In this section, an inventory model which consists of two independent inventory systems as described in the previous section is considered. For simplicity of discussion, both of them are assumed to be identical. A special feature of this model is that lateral transshipment of returns between the inventory systems is allowed. Lateral transshipment of demands has been studied in a number of papers [49, 76]. Substantial savings can be realized by sharing of inventory via the lateral transshipment of demands [179]. Here, this concept is extended to the handling of returns. Recall that an arrived return will be disposed if the inventory level is Q in the previous model. In the new model, lateral transshipment of returns between the inventory systems is allowed whenever one of them is full (whenever the inventory level is Q) and the other is not yet full (the inventory level is less than Q). Denote $x(t)$ and $y(t)$ to be the inventory levels of the serviceable product in the first and the second inventory system at time t respectively. Then, the random variables $x(t)$ and $y(t)$ take integral values in $[0, Q]$. Thus, the joint inventory process

$$\{(x(t), y(t)), \ t \geq 0\}$$

is again a continuous time Markov chain taking values in the state space

$$S = \{(x, y) : x = 0, \cdots, Q, \quad y = 0, \cdots, Q.\}.$$

The inventory states were ordered lexicographically, according to x first and then y. The generator matrix for the joint inventory system can be written by using Kronecker tensor product as follows:

$$B = I_{Q+1} \otimes A + A \otimes I_{Q+1} + \Delta \otimes \Lambda + \Lambda \otimes \Delta \tag{3.5}$$

where

$$\Lambda = \begin{pmatrix} 1 & & & & 0 \\ -1 & 1 & & & \\ & \ddots & \ddots & & \\ & & -1 & 1 & \\ 0 & & & -1 & 0 \end{pmatrix} \tag{3.6}$$

and

$$\Delta = \begin{pmatrix} 0 & & & & 0 \\ & 0 & & & \\ & & \ddots & & \\ & & & 0 & \\ 0 & & & & a\mu \end{pmatrix} \tag{3.7}$$

and I_{Q+1} is the $(Q+1) \times (Q+1)$ identity matrix. The steady state probability vector $\mathbf{q}$ satisfies

$$B\mathbf{q} = \mathbf{0} \quad \text{and} \quad \mathbf{1}^T \mathbf{q} = 1. \tag{3.8}$$

We note that the generator B is irreducible and it has a one-dimensional null-space with a right positive null vector, see [101, 203]. The steady state probability vector $\mathbf{q}$ is the normalized form of the positive null vector of B. Let q_{ij} be the steady state probability that the inventory level of the serviceable product is i in the first inventory system and j in the second inventory system. Many important quantities of the system performance can be written in terms of q_{ij}. For example the return rejection probability is q_{QQ}. Unfortunately, closed form solution of $\mathbf{q}$ is not generally available. Very often by making use of the block structure of the generator matrix B, classical iterative methods such as Block Gauss-Seidel (BGS) method is applied to solve the steady state probability distribution [50, 101, 203]. In the following, instead of solving the steady state probability distribution numerically, closed form approximation for the probability distribution $\mathbf{q}$ is derived under some assumptions.

Proposition 3.5. *Let $\mathbf{p}$ be the steady state probability distribution for the generator matrix A in Proposition 3.1. If $\rho < 1$ then*

$$\|B(\mathbf{p} \otimes \mathbf{p})\|_\infty \leq \frac{4a\mu}{(Q+1)^2(1-\rho)^2}$$

The probability vector $\mathbf{q} = \mathbf{p} \otimes \mathbf{p}$ is an approximation of the steady state probability vector when Q is large.

Proof. The probability vector $\mathbf{p}$ is just the solution of (3.2). By direct verification, one have $\mathbf{1}^t(\mathbf{p} \otimes \mathbf{p}) = 1$ and

$$(I \otimes A + A \otimes I)(\mathbf{p} \otimes \mathbf{p}) = (\mathbf{p} \otimes A\mathbf{p} + A\mathbf{p} \otimes \mathbf{p}) = (\mathbf{p} \otimes \mathbf{0} + \mathbf{0} \otimes \mathbf{p}) = \mathbf{0}.$$

Therefore from (3.5)

$$B(\mathbf{p} \otimes \mathbf{p}) = (\Lambda \otimes \Delta)(\mathbf{p} \otimes \mathbf{p}) + (\Delta \otimes \Lambda)(\mathbf{p} \otimes \mathbf{p}) = (\Lambda\mathbf{p} \otimes \Delta\mathbf{p}) + (\Delta\mathbf{p} \otimes \Lambda\mathbf{p}).$$

One could observe that

$$||\Lambda||_\infty = 2, \quad ||\mathbf{p}||_\infty \le K \quad \text{and} \quad ||\Delta||_\infty = a\mu.$$

The l_∞-norm of an $p \times q$ matrix Z is defined as follows:

$$||Z||_\infty = \max\left\{\sum_{j=1}^{q} |Z_{1j}|, \sum_{j=1}^{q} |Z_{2j}|, \cdots, \sum_{j=1}^{q} |Z_{pj}|\right\}.$$

Therefore,

$$\begin{aligned}
||B(\mathbf{p} \otimes \mathbf{p})||_\infty &\le 2||\Lambda||_\infty||\mathbf{p}||_\infty||\Delta||_\infty||\mathbf{p}||_\infty \\
&= 4a\mu K^2 \\
&\le \frac{4a\mu}{(Q+1)^2(1-\rho)^2}
\end{aligned} \tag{3.9}$$

If one adopt $\mathbf{q} = \mathbf{p} \otimes \mathbf{p}$ to be the system steady state probability distribution, then the approximated optimal replenishment size of each inventory system is the same as in Proposition 3.3. By allowing transshipment of returns, the rejection rate of returns of the two inventory systems will be decreased from

$$2\mu K(1 - \rho^{Q+1}) \approx \frac{2\mu}{Q+1}$$

to

$$\mu K^2(1 - \rho^{Q+1})^2 \approx \frac{\mu}{(Q+1)^2}.$$

Note that the approximation is valid only if Q is large, the error is of order $O(Q^{-2})$.

3.4 The Hybrid Re-manufacturing Systems

In this section, we propose a hybrid system, a system consists of a re-manufacturing process and a manufacturing process. The proposed hybrid system captures the re-manufacturing process and the system can produce serviceable product when the return rate is zero. The demands and the returns are assumed to follow independent Poisson processes. The serviceable product inventory level and the outside procurements are controlled by a popular (r, Q) continuous review policy. The inventory level of the serviceable product is modelled as an irreducible continuous time Markov chain and the generator matrix is constructed. It is found that the generator matrix has a near-Toeplitz structure.

Then a direct method is proposed for solving the steady state probabilities. The direct method is based on Fast Fourier Transforms (FFTs) and the Sherman-Morrison-Woodbury Formula (Proposition 1.36). The complexity of the method is then given and some special cases analysis are also discussed.

3.4.1 The Hybrid System

In this subsection, an inventory model which captures the re-manufacturing process is proposed. Disposal of returned product is allowed when the return capacity is full. In the model, there are two types of inventory to be managed, the serviceable product and the returned product. The demands and the returns are assumed to follow independent Poisson process with mean rates λ and γ respectively. The re-manufacturing process is then modelled by an $M/M/1/N$ queue: a returned product acts as a customer and a reliable re-manufacturing machine (with processing rate μ) acts as the server in the queue. The re-manufacturing process is stopped whenever there is no space for placing the serviceable product (ie. when the serviceable product inventory level is Q). Here we also assume that when the return level is zero, the system can produce at a rate of τ (exponentially distributed).

The serviceable product inventory level and the outside procurements are controlled by a popular (r, Q) continuous review policy. This means that when the inventory level drops to r, an outside procurement order of size $(Q - r)$ is placed and arrived at once. For simplicity of discussion, the procurement level r is assumed to be -1. This means that whenever there is no serviceable product in the system and there is an arrival of demand then a procurement order of size $(Q + 1)$ is placed and arrived at once. Therefore the procurement can clear the backlogged demand and bring the serviceable product inventory to Q. We also assume that it is always possible to purchase the required procurement. The inventory levels of both the returns and the serviceable product are modelled as Markovian process. The capacity N for the returns and the capacity Q for serviceable product Q are assumed to be large. Fig. 3.3 (Taken from [73, 77]) gives the framework of the re-manufacturing system.

3.4.2 The Generator Matrix of the System

In this subsection, the generator matrix for the re-manufacturing system is constructed. Let $x(t)$ and $y(t)$ be the inventory levels of the returns and the inventory levels of the serviceable products at time t respectively. Then $x(t)$ and $y(t)$ take integral values in $[0, N]$ and $[0, Q]$ respectively. The joint inventory process

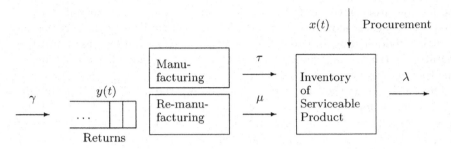

Fig. 3.3. The hybrid system.

$$\{(x(t), y(t)),\ t \geq 0\}$$

is a continuous time Markov chain taking values in the state space

$$S = \{(x, y) : x = 0, \cdots, N, \quad y = 0, \cdots, Q\}.$$

By ordering the joint inventory states lexicographically, according to x first and then y, the generator matrix for the joint inventory system can be written as follows:

$$A_1 = \begin{pmatrix} B_0 & -U & & & 0 \\ -\gamma I_{Q+1} & B & -U & & \\ & \ddots & \ddots & \ddots & \\ & & -\gamma I_{Q+1} & B & -U \\ 0 & & & -\gamma I_{Q+1} & B_N \end{pmatrix}, \tag{3.10}$$

where

$$U = \begin{pmatrix} 0 & & & & 0 \\ \mu & 0 & & & \\ & \ddots & \ddots & & \\ & & & \ddots & \ddots \\ 0 & & & & \mu & 0 \end{pmatrix}, \tag{3.11}$$

$$B_0 = \gamma I_{Q+1} + \begin{pmatrix} \tau + \lambda & -\lambda & & & 0 \\ -\tau & \tau + \lambda & -\lambda & & \\ & -\tau & \ddots & & -\lambda \\ & & & \ddots & \tau + \lambda & -\lambda \\ -\lambda & & & & -\tau & \lambda \end{pmatrix}, \tag{3.12}$$

$$B = \gamma I_{Q+1} + \begin{pmatrix} \lambda + \mu & -\lambda & & & 0 \\ & \lambda + \mu & -\lambda & & \\ & & \ddots & -\lambda & \\ & & & \lambda + \mu & -\lambda \\ -\lambda & & & & \lambda \end{pmatrix}, \tag{3.13}$$

$$B_N = B - \gamma I_{Q+1}.$$

Here I_{Q+1} is the $(Q+1) \times (Q+1)$ identity matrix. The steady state probability distribution $\mathbf{p}$ is required if one wants to get the performance of the system. Note that the generator A_1 is irreducible and from the Perron and Frobenius theory [101] it is known that it has a 1-dimensional null-space with a right positive null vector. Hence, as mentioned in Section 3.2.1, one can consider an equivalent linear system instead.

$$G\mathbf{x} \equiv (A_1 + \mathbf{ff}^T)\mathbf{x} = \mathbf{f}, \quad \text{where} \quad \mathbf{f} = (0, \ldots, 0, 1)^T. \tag{3.14}$$

Proposition 3.6. *The matrix G is nonsingular.*

However, the closed form solution of $\mathbf{p}$ is not generally available. Iterative methods such as (PCG) method is efficient in solving the probability vector $\mathbf{p}$ when one of the parameters N and Q is fixed, see for instance [48, 50, 52, 55]. However, when both Q and N are getting large, the fast convergence rate of PCG method cannot be guaranteed especially when the smallest singular value tends to zero very fast [49, 53]. Other approximation methods for solving the problem can be found in [50]. In the following subsection, a direct method is proposed for solving (3.14).

3.4.3 The Direct Method

We consider taking circulant approximations to the matrix blocks in A_1. We define the following circulant matrices:

$$c(G) = \begin{pmatrix} c(\bar{B}_0) & -c(U) & & & \\ -\gamma I_{Q+1} & c(B) & -c(U) & & \\ & \ddots & \ddots & \ddots & \\ & & -\gamma I_{Q+1} & c(B) & -c(U) \\ & & & -\gamma I_{Q+1} & c(B_N) \end{pmatrix}, \tag{3.15}$$

where

$$c(U) = \begin{pmatrix} 0 & & & & \mu \\ \mu & 0 & & & \\ & \ddots & \ddots & & \\ & & & \ddots & \ddots \\ 0 & & & \mu & 0 \end{pmatrix}, \tag{3.16}$$

$$\tag{3.17}$$

$$c(\bar{B}_0) = \gamma I_{Q+1} + \begin{pmatrix} \tau+\lambda & -\lambda & & & -\tau \\ -\tau & \tau+\lambda & -\lambda & & \\ & \ddots & \ddots & -\lambda & \\ & & \ddots & \tau+\lambda & -\lambda \\ -\lambda & & & -\tau & \tau+\lambda \end{pmatrix}, \qquad (3.18)$$

$$c(B) = \gamma I_{Q+1} + \begin{pmatrix} \lambda+\mu & -\lambda & & & 0 \\ & \lambda+\mu & -\lambda & & \\ & & \ddots & -\lambda & \\ & & & \lambda+\mu & -\lambda \\ -\lambda & & & & \lambda+\mu \end{pmatrix}, \qquad (3.19)$$

$$(3.20)$$

$$c(B_N) = c(B) - \gamma I_{Q+1}. \qquad (3.21)$$

We observe that

$$c(U) - U = \mu e_1^T e_{Q+1}, \quad c(\bar{B}_0) - \bar{B}_0 = -\tau e_1^T e_{Q+1},$$

$$c(B) - B = \mu e_{Q+1}^T e_{Q+1}, \quad \text{and} \quad c(B_N) - B_N = \mu e_{Q+1}^T e_{Q+1}$$

where

$$\mathbf{e}_1 = (1, 0, \ldots, 0) \quad \text{and} \quad \mathbf{e}_{Q+1} = (0, \ldots, 0, 1)$$

are 1-by-$(Q+1)$ unit vectors. Here we remark that

$$\bar{B}_0 = B_0 + \tau e_{Q+1}^T e_{Q+1}.$$

Therefore the matrix G is a sum of a circulant block matrix and another block matrix with small rank except the first and the last diagonal blocks.

In view of the above formulation, the problem is equivalent to consider the solution of the linear system having the form $A\mathbf{z} = \mathbf{b}$ where A is a block-Toeplitz matrix given by

$$A = \begin{pmatrix} A_{11} & \cdots\cdots & A_{1m} \\ A_{21} & \cdots\cdots & A_{2m} \\ \vdots & \vdots \quad \vdots & \vdots \\ A_{m1} & \cdots\cdots & A_{mm} \end{pmatrix}. \qquad (3.22)$$

Here

$$A_{ij} = C_{i-j} + \mathbf{u}_{i-j}^T \mathbf{v} \qquad (3.23)$$

where C_{i-j} is an $n \times n$ circulant matrix, and $\mathbf{u}_{i-j}$ and $\mathbf{v}$ are $k \times n$ matrices and $k << m, n$ so that A_{ij} is an $n \times n$ near-circulant matrix, i.e., finite rank being less than or equal to k. We remark that the class of matrices A is closely related to the generator matrices of many Markovian models such as queueing systems [50, 142, 143], manufacturing systems [48, 50, 52, 55, 58] and re-manufacturing systems [76, 92, 201].

Next, we note that an $n \times n$ circulant matrix can be diagonalized by using the discrete Fourier matrix F_n. Moreover, its eigenvalues can be obtained in $O(n \log n)$ operations by using the FFT, see for instance Davis [82]. In view of this advantage, consider

$$(I_m \otimes F_n^*)A(I_m \otimes F_n) = \begin{pmatrix} D_{11} & \cdots & D_{1m} \\ D_{21} & \cdots & D_{2m} \\ \vdots & \vdots & \vdots \\ D_{m1} & \cdots & D_{mm} \end{pmatrix} + \begin{pmatrix} E_{11} & \cdots & E_{1m} \\ E_{21} & \cdots & E_{2m} \\ \vdots & \vdots & \vdots \\ E_{m1} & \cdots & E_{mm} \end{pmatrix} \tag{3.24}$$
$$\equiv D + E.$$

Here D_{ij} is a diagonal matrix containing the eigenvalues of C_{i-j} and

$$E_{ij} = (F_n^* \mathbf{u}_{i-j}^T)(\mathbf{v} F_n) \equiv (\mathbf{x}_{i-j}^T)(\mathbf{y}). \tag{3.25}$$

Also note that

$$E = \begin{pmatrix} \mathbf{x}_0^T \mathbf{y} & \cdots\cdots & \mathbf{x}_{1-m}^T \mathbf{y} \\ \mathbf{x}_1^T \mathbf{y} & \cdots\cdots & \mathbf{x}_{2-m}^T \mathbf{y} \\ \vdots & \vdots & \vdots \\ \mathbf{x}_{m-1}^T \mathbf{y} & \cdots\cdots & \mathbf{x}_0^T \mathbf{y} \end{pmatrix}$$
$$= \begin{pmatrix} \mathbf{x}_0^T & \cdots & \mathbf{x}_{1-m}^T \\ \mathbf{x}_1^T & \cdots & \mathbf{x}_{2-m}^T \\ \vdots & \vdots & \vdots \\ \mathbf{x}_{m-1}^T & \cdots & \mathbf{x}_0^T \end{pmatrix} \begin{pmatrix} \mathbf{y} & 0 & \cdots & 0 & 0 \\ 0 & \mathbf{y} & 0 & \cdots & 0 \\ \vdots & \vdots & \ddots & \ddots & \vdots \\ 0 & \cdots\cdots & 0 & \mathbf{y} \end{pmatrix} \tag{3.26}$$
$$\equiv XY.$$

Note that D is still a block-Toeplitz matrix and there exists a permutation matrix P such that

$$PDP^T = \mathrm{diag}(T_1, T_2, \ldots, T_n) \tag{3.27}$$

where T_i is an $m \times m$ Toeplitz matrix. In fact direct methods for solving Toeplitz systems that are based on the recursion formula are in constant use, see for instance, Trench [199]. For an $m \times m$ Toeplitz matrix T_i, these methods require $O(m^2)$ operations. Faster algorithms that require $O(m \log^2 m)$ operations have been developed for symmetric positive definite Toeplitz matrices, see Ammar and Gragg [5] for instance. The stability properties of these direct methods are discussed in Bunch [38]. Hence by using direct methods, the linear system $D\mathbf{z} = \mathbf{b}$ can be solved in $O(nm^2)$ operations. The matrix X is an $mn \times mk$ matrix and the matrix Y is an $mk \times mn$ matrix.

To solve the linear system, we apply the Sherman-Morrison-Woodbury Formula (Proposition 1.36). The solution of $A\mathbf{z} = \mathbf{b}$ can be written as follows:

$$\mathbf{z} = D^{-1}\mathbf{b} - D^{-1}X(I_{mk} + YD^{-1}X)^{-1}YD^{-1}\mathbf{b}. \tag{3.28}$$

3.4.4 The Computational Cost

In this section, the computational cost of the proposed method is discussed. The main computational cost of (3.28) consists of

(C0) FFT operations in (3.25);
(C1) Solving $\mathbf{r} = D^{-1}\mathbf{b}$;
(C2) Solving $W = D^{-1}X$;
(C3) Matrix multiplication of YW;
(C4) Matrix multiplication of $Y\mathbf{r}$;
(C5) Solving $(I_{mk} + YD^{-1}X)^{-1}\mathbf{r}$.

The operational cost for (C0) is of $O(mn \log n)$. The operational cost for (C1) is at most $O(nm^2)$ operations by using direct solvers for Toeplitz system. The cost for (C2) is at most $O(knm^3)$ operations in view of (C1). The operational cost for (C3) is of $O(k^2nm^2)$ because of the sparse structure of Y. The cost for (C4) is $O(knm)$ operations. Finally the cost of (C5) is $O((km)^3)$ operations. Hence the overall cost will be $(km^3(n + k^2))$ operations.

In fact the nice structure of D allows us to solve $D\mathbf{r} = \mathbf{b}$ in a parallel computer. Moreover $DW = X$ consists of n separate linear systems (a multiple right hand sides problem). Again this can also be solved in a parallel computer. Therefore the cost of (C1) and (C2) can be reduced by using parallel algorithms. Assuming that k is small, the costs of (C1) and (C2) can be reduced to $O(m^2)$ and $(O(m^3))$ operations time units respectively when n parallel processors are used.

3.4.5 Some Special Cases Analysis

In this section, k is assumed to be small and some special cases of solving (3.28) is discussed.

Case (i) When all the $\mathbf{u}_{i-j}$ in (3.23) are equal, then we see that all the columns of X are equal and the cost (C2) will be at most $O(nm^2)$ operations. Hence the overall cost will be $O(m^2(m + n) + mn \log n)$ operations.

Case (ii) If the matrix A is a block-circulant matrix, then all the matrices T_i in (3.27) are circulant matrices. The cost of (C1) and (C2) can be reduced to $O(nm \log m)$ and $O(nm^2 \log m)$ operations respectively. Hence the overall cost will be $O(m^3 + nm(m \log m + \log n))$ operations.

Case (iii) If the matrix A is a block tri-diagonal matrix, then all the matrices T_i in (3.27) are tri-diagonal matrices. The cost of (C0) will be $O(n \log n)$. The cost of (C1) and (C2) can be reduced to $O(nm)$ and $O(nm^2)$ operations respectively. Hence the overall cost will be $O(m^3 + n(m^2 + \log n))$ operations.

We end this section by the following proposition. The proposition gives the complexity for solving the steady state probability distribution $\mathbf{p}$ for the generator matrix (3.10) when $Q \approx N$.

Proposition 3.7. *The steady state probability distribution* $\mathbf{p}$ *can be obtained in* $O(N^3)$ *operations when* $Q \approx N$.

Proof. In the view of case (iii) in this section, the complexity of our method for solving (3.14) is $O(N^3)$ when $Q \approx N$ while the complexity of solving (3.14) by LU decomposition is $O(N^4)$.

3.5 Summary

In this chapter, we present the concept of re-manufacturing systems. Several stochastic models for re-manufacturing systems are discussed. The steady state probability distributions of the models are either obtained in closed form or can be solved by fast numerical algorithms. The models here concern only single-item, it will be interesting to extend the results to multi-item case.

4

Hidden Markov Model for Customers Classification

4.1 Introduction

In this chapter, a new simple Hidden Markov Model (HMM) is proposed. The process of the proposed HMM can be explained by the following example.

4.1.1 A Simple Example

We consider the process of choosing a die of four faces (a tetrahedron) and recording the number of dots obtained by throwing the die [173]. Suppose we have two dice A and B, each of them has four faces $(1, 2, 3$ and $4)$. Moreover, Die A is fair and Die B is biased. The probability distributions of dots obtained by throwing dice A and B are given in Table 4.1.

Table 4.1. Probability distributions of dice A and dice B.

Dice	1	2	3	4
A	1/4	1/4	1/4	1/4
B	1/6	1/6	1/3	1/3

Each time a die is to be chosen, we assume that with probability α, Die A is chosen, and with probability $(1-\alpha)$, Die B is chosen. This process is hidden as we don't know which die is chosen. The value of α is to be determined. The chosen die is then thrown and the number of dots (this is observable) obtained is recorded. The following is a possible realization of the whole process:

$$A \rightarrow 1 \rightarrow A \rightarrow 2 \rightarrow B \rightarrow 3 \rightarrow A \rightarrow 4 \rightarrow B \rightarrow 1 \rightarrow B \rightarrow 2 \rightarrow \cdots \rightarrow .$$

We note that the whole process of the HMM can be modelled by a classical Markov chain model with the transition probability matrix being given by

$$\begin{array}{c} A \\ B \\ 1 \\ 2 \\ 3 \\ 4 \end{array} \left(\begin{array}{cc|cccc} 0 & 0 & \alpha & \alpha & \alpha & \alpha \\ 0 & 0 & 1-\alpha & 1-\alpha & 1-\alpha & 1-\alpha \\ \hline 1/4 & 1/6 & 0 & 0 & 0 & 0 \\ 1/4 & 1/6 & 0 & 0 & 0 & 0 \\ 1/4 & 1/3 & 0 & 0 & 0 & 0 \\ 1/4 & 1/3 & 0 & 0 & 0 & 0 \end{array} \right).$$

The rest of the chapter is organized as follows. In Section 4.2, the estimation method will be demonstrated by the example giving in Section 4.1. In Section 4.3, the proposed method is extended to a general case. In Section 4.4, some analytic results of a special case are presented. In Section 4.5, an application in customers classification with practical data taken from a computer service company is presented and analyzed. Finally, a brief summary is given in Section 4.6 to conclude this chapter.

4.2 Parameter Estimation

In this section, we introduce a simple estimation method of α, Ching and Ng [60] Clearly in order to define the HMM, one has to estimate α from an observed data sequence. We suppose that the distribution of dots (in steady state) is given by

$$(\frac{1}{6}, \frac{1}{4}, \frac{1}{4}, \frac{1}{3})^T$$

then the question is: how to estimate α? We note that

$$P^2 = \left(\begin{array}{cc|cccc} \alpha & \alpha & 0 & 0 & 0 & 0 \\ 1-\alpha & 1-\alpha & 0 & 0 & 0 & 0 \\ \hline 0 & 0 & \frac{1}{6}+\frac{\alpha}{12} & \frac{1}{6}+\frac{\alpha}{12} & \frac{1}{6}+\frac{\alpha}{12} & \frac{1}{6}+\frac{\alpha}{12} \\ 0 & 0 & \frac{1}{6}+\frac{\alpha}{12} & \frac{1}{6}+\frac{\alpha}{12} & \frac{1}{6}+\frac{\alpha}{12} & \frac{1}{6}+\frac{\alpha}{12} \\ 0 & 0 & \frac{1}{3}-\frac{\alpha}{12} & \frac{1}{3}-\frac{\alpha}{12} & \frac{1}{3}-\frac{\alpha}{12} & \frac{1}{3}-\frac{\alpha}{12} \\ 0 & 0 & \frac{1}{3}-\frac{\alpha}{12} & \frac{1}{3}-\frac{\alpha}{12} & \frac{1}{3}-\frac{\alpha}{12} & \frac{1}{3}-\frac{\alpha}{12} \end{array} \right) = \left(\begin{array}{cc} R & 0 \\ 0 & \tilde{P} \end{array} \right).$$

If we ignore the hidden states (the first diagonal block R), then the observable states follow the transition probability matrix given by the following matrix

$$\tilde{P} = \left(\begin{array}{cccc} \frac{1}{6}+\frac{\alpha}{12} & \frac{1}{6}+\frac{\alpha}{12} & \frac{1}{6}+\frac{\alpha}{12} & \frac{1}{6}+\frac{\alpha}{12} \\ \frac{1}{6}+\frac{\alpha}{12} & \frac{1}{6}+\frac{\alpha}{12} & \frac{1}{6}+\frac{\alpha}{12} & \frac{1}{6}+\frac{\alpha}{12} \\ \frac{1}{3}-\frac{\alpha}{12} & \frac{1}{3}-\frac{\alpha}{12} & \frac{1}{3}-\frac{\alpha}{12} & \frac{1}{3}-\frac{\alpha}{12} \\ \frac{1}{3}-\frac{\alpha}{12} & \frac{1}{3}-\frac{\alpha}{12} & \frac{1}{3}-\frac{\alpha}{12} & \frac{1}{3}-\frac{\alpha}{12} \end{array} \right)$$

$$= \left(\begin{array}{c} \frac{1}{6}+\frac{\alpha}{12} \\ \frac{1}{6}+\frac{\alpha}{12} \\ \frac{1}{3}-\frac{\alpha}{12} \\ \frac{1}{3}-\frac{\alpha}{12} \end{array} \right) (1,1,1,1).$$

Thus it is easy to see that the stationary probability distribution of $\tilde{P}$ is given by

$$\mathbf{p} = (\frac{1}{6} + \frac{\alpha}{12}, \frac{1}{6} + \frac{\alpha}{12}, \frac{1}{3} - \frac{\alpha}{12}, \frac{1}{3} - \frac{\alpha}{12})^T.$$

This probability distribution $\mathbf{p}$ should be consistent with the observed distribution $\mathbf{q}$ of the observed sequence, i.e.

$$\mathbf{p} = (\frac{1}{6} + \frac{\alpha}{12}, \frac{1}{6} + \frac{\alpha}{12}, \frac{1}{3} - \frac{\alpha}{12}, \frac{1}{3} - \frac{\alpha}{12})^T \approx \mathbf{q} = (\frac{1}{6}, \frac{1}{4}, \frac{1}{4}, \frac{1}{3})^T.$$

This suggests a nature method to estimate α. The unknown transition probability α can then be obtained by solving the minimisation problem:

$$\min_{0 \leq \alpha \leq 1} ||\mathbf{p} - \mathbf{q}||.$$

If we choose $||.||$ to be the $||.||_2$ then one may consider the following minimisation problem:

$$\min_{0 \leq \alpha \leq 1} ||\mathbf{p} - \mathbf{q}||_2^2 = \min_{0 \leq \alpha \leq 1} \sum_{i=1}^{4} (\mathbf{p}_i - \mathbf{q}_i)^2.$$

In this case, it is a standard constrained least squares problem and can be solved easily. For more detailed discussion on statistical inference of a HMM, we refer readers to the book by MacDonald and Zucchini [149].

4.3 Extension of the Method

In this section, the parameter estimation method is extended to a general HMM with m hidden states and n observable states. In general the number of hidden states can be more than two. Suppose the number of hidden states is m and the stationary distribution of the hidden states is given by

$$\alpha = (\alpha_1, \alpha_2, \ldots, \alpha_m).$$

Suppose the number of observable state is n and when the hidden state is $i (i = 1, 2, \ldots, m)$, the stationary distribution of the observable states is

$$(p_{i1}, p_{i2}, \ldots, p_{in}).$$

We assume that m, n and p_{ij} are known. Given an observed sequence of the observable states, one can calculate the occurrences of each state in the sequence and hence the observed distribution $\mathbf{q}$. Using the same trick discussed in Section 3, if we ignore the hidden states, the observable states follow the one-step transition probability matrix:

$$\tilde{P}_2 = \begin{pmatrix} p_{11} & p_{21} & \cdots & p_{m1} \\ p_{12} & p_{22} & \cdots & p_{m2} \\ \vdots & \vdots & \vdots & \vdots \\ p_{1n} & p_{2n} & \cdots & p_{mn} \end{pmatrix} \begin{pmatrix} \alpha_1 & \alpha_2 & \cdots & \alpha_1 \\ \alpha_2 & \alpha_2 & \cdots & \alpha_2 \\ \vdots & \vdots & \vdots & \vdots \\ \alpha_m & \alpha_m & \cdots & \alpha_m \end{pmatrix} = \mathbf{p}(1,1,\ldots,1) \qquad (4.1)$$

where

$$\mathbf{p} = \left(\sum_{k=1}^{m} \alpha_k p_{k1}, \sum_{k=1}^{m} \alpha_k p_{k2}, \ldots, \sum_{k=1}^{m} \alpha_k p_{kn} \right)^T.$$

It is easy to check that

$$\tilde{P}_2 \mathbf{p} = \mathbf{p} \quad \text{and} \quad \sum_{k=1}^{n} \mathbf{p}_k = 1.$$

Thus the following proposition can be proved easily.

Proposition 4.1. *The vector* $\mathbf{p}$ *is the stationary probability distribution of* $\tilde{P}_2$.

Therefore the transition probabilities of the hidden states

$$\alpha = (\alpha_1, \alpha_2, \ldots, \alpha_m)$$

can be obtained by solving

$$\min_{\alpha} \|\mathbf{p} - \mathbf{q}\|_2^2$$

subject to

$$\sum_{k=1}^{m} \alpha_k = 1 \quad \text{and} \quad \alpha_k \geq 0.$$

4.4 Special Case Analysis

In this section, a detailed discussion is given for the model having 2 hidden states. In this case one may re-write (4.1) as follows:

$$\tilde{P} = \begin{pmatrix} p_{11} & p_{21} \\ p_{12} & p_{22} \\ \vdots & \vdots \\ p_{1n} & p_{2n} \end{pmatrix} \begin{pmatrix} \alpha_1 & \alpha_1 & \cdots & \alpha_1 \\ 1-\alpha_1 & 1-\alpha_1 & \cdots & 1-\alpha_1 \end{pmatrix} = \mathbf{p}(1,1,\ldots,1) \qquad (4.2)$$

where

$$\mathbf{p} = (\alpha p_{11} + (1-\alpha)p_{21}, \alpha p_{12} + (1-\alpha)p_{22}, \ldots, \alpha p_{1n} + (1-\alpha)p_{2n})^T.$$

It is easy to check that

$$\bar{P}\mathbf{p} = \mathbf{p} \quad \text{and} \quad \sum_{i=1}^{n} \mathbf{p}_i = 1$$

and therefore $\mathbf{p}$ is the steady state probability distribution.

Suppose the observed distribution $\mathbf{q}$ of the observable states is given, then α can be estimated by the following minimization problem:

$$\min_{\alpha} ||\mathbf{p} - \mathbf{q}||_2^2$$

subject to $0 \leq \alpha \leq 1$ or equivalently

$$\min_{0 \leq \alpha \leq 1} \sum_{k=1}^{n} \{\alpha p_{1k} + (1 - \alpha)p_{2k} - q_k\}^2.$$

The following proposition can be obtained by direct verification.

Proposition 4.2. *Let*

$$\tau = \frac{\sum_{j=1}^{n}(q_j - p_{2j})(p_{1j} - p_{2j})}{\sum_{j=1}^{n}(p_{1j} - p_{2j})^2}$$

then the optimal value of α is given as follows:

$$\alpha = \begin{cases} 0 \text{ if } \tau \leq 0; \\ \tau \text{ if } 0 < \tau < 1; \\ 1 \text{ if } \tau \geq 1. \end{cases}$$

One may interpret the result in Proposition 4.2 as follows.

$$\tau = \frac{< (\mathbf{q} - \mathbf{p}_2), (\mathbf{p}_1 - \mathbf{p}_2) >}{< (\mathbf{p}_1 - \mathbf{p}_2), (\mathbf{p}_1 - \mathbf{p}_2) >} = \frac{||\mathbf{q} - \mathbf{p}_2||_2 \cos(\theta)}{||\mathbf{p}_1 - \mathbf{p}_2||_2}. \quad (4.3)$$

Here $< .,. >$ is the standard inner product on the vector space R^n,

$$\mathbf{p}_1 = (p_{11}, p_{12}, \ldots, p_{1n})^T$$

and

$$\mathbf{p}_2 = (p_{21}, p_{22}, \ldots, p_{2n})^T.$$

Moreover, $||.||_2$ is the L_2-norm on R^n and θ is the angle between the vectors

$$(\mathbf{q} - \mathbf{p}_2) \quad \text{and} \quad (\mathbf{p}_1 - \mathbf{p}_2).$$

Two hyperplanes H_1 and H_2 are defined in R^n. Both hyperplanes are perpendicular to the vector $(\mathbf{p}_1 - \mathbf{p}_2)$ and H_i contains the point $\mathbf{p}_i$ (distribution) for

$i = 1, 2$, see Fig. 4.1 (Taken from [69]). From (4.3), Proposition 4.2 and Fig. 4.4, any point $\mathbf{q}'$ on the left of the hyperplane H_1 has the following property:

$$||\mathbf{q} - \mathbf{p}_2||_2 \cos(\theta) \geq ||\mathbf{p}_1 - \mathbf{p}_2||_2.$$

Hence for such $\mathbf{q}'$, the optimal α is 1. For a point $\mathbf{q}''$ on the right of the hyperplane H_2, then $\cos(\theta) \leq 0$ and hence the optimal α is zero. Lastly, for a point $\mathbf{q}$ in between the two hyperplanes, the optimal α lies between 0 and 1 and the optimal value is given by τ in (4.3). This special case motivates us to apply the HMM in the classification of customers.

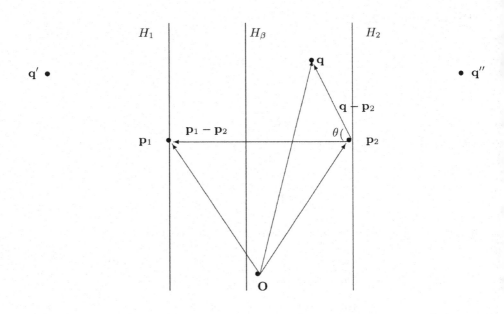

Fig. 4.1. The graphical interpretation of Proposition 4.2.

4.5 Application to Classification of Customers

In this section, the HMM discussed in the Section 4.4 is applied to the customers classification of a computer service company. We remark that there are a number of classification methods such as machine learning and Bayesian learning, interested readers can consult the book by Young and Calvert [214]. In this problem, HMM is an efficient and effective classification method but we make no claim that HMM is the best one.

A computer service company offers four types of distant calls services I, II, III and IV (four different periods of a day). From the customer database of the users, the information of the expenditure distribution of 71 randomly chosen customers is obtained. A longitudinal study has been carried out for half a year to investigate the customers. Customers' behavior and responses are captured and monitored during the period of investigation. For simplicity of discussion, the customers are classified into two groups. Among them 22 customers are known to be loyal customers (Group A) and the other 49 customers are not loyal customers (Group B). This classification is useful to marketing managers when they plan any promotions. For the customers in Group A, promotions on new services and products will be given to them. While for the customers in Group B, discount on the current services will be offered to them to prevent them from switching/churning to the competitor companies.

Two-third of the data are used to build the HMM and the remaining data are used to validate the model. Therefore, 16 candidates are randomly taken (these customers are labelled in the first 16 customers in Table 4.2) from Group A and 37 candidates from group B. The remaining 6 candidates (the first 6 customers in Table 4.2) from Group A and 12 candidates from Group B are used for validating the constructed HMM. A HMM having four observable states (I, II, III and IV) and two hidden states (Group A and Group B) is then built.

From the information of the customers in Group A and Group B in Table 4.3, the average expenditure distributions for both groups are computed in Table 4.3. This means that a customer in Group A (Group B) is characterized by the expenditure distribution in the first (second) row of Table 4.3.

An interesting problem is the following. Given the expenditure distribution of a customer, how to classify the customer correctly (Group A or Group B) based on the information in Table 4.4? To tackle this problem, one can apply the method discussed in previous section to compute the transition probability α in the hidden states. This value of α can be used to classify a customer. If α is close to 1 then the customer is likely to be a loyal customer. If α is close to 0 then the customer is likely to be a not-loyal customer.

The values of α for all the 53 customers are listed in Table 4.2. It is interesting to note that the values of α of all the first 16 customers (Group A) lie in the interval $[0.83, 1.00]$. While the values of α of all the other customers (Group B) lie in the interval $[0.00, 0.69]$. Based on the values of α obtained, the two groups of customers can be clearly separated by setting the cutoff value β to be 0.75. A possible decision rule can therefore be defined as follows: Classify a customer to Group A if $\alpha \geq \beta$, otherwise classify the customer to Group B. Referring to Fig. 4.1, it is clear that the customers are separated by the

Table 4.2. Two-third of the data are used to build the HMM.

Customer	I	II	III	IV	α	Customer	I	II	III	IV	α
1	1.00	0.00	0.00	0.00	1.00	2	1.00	0.00	0.00	0.00	1.00
3	0.99	0.01	0.00	0.00	1.00	4	0.97	0.03	0.00	0.00	1.00
5	0.87	0.06	0.04	0.03	0.98	6	0.85	0.15	0.00	0.00	0.92
7	0.79	0.18	0.02	0.01	0.86	8	0.77	0.00	0.23	0.00	0.91
9	0.96	0.01	0.00	0.03	1.00	10	0.95	0.00	0.02	0.03	1.00
11	0.92	0.08	0.00	0.00	1.00	12	0.91	0.09	0.00	0.00	1.00
13	0.83	0.00	0.17	0.00	0.97	14	0.82	0.18	0.00	0.00	0.88
15	0.76	0.04	0.00	0.20	0.87	16	0.70	0.00	0.00	0.30	0.83
17	0.62	0.15	0.15	0.08	0.69	18	0.57	0.14	0.00	0.29	0.62
19	0.56	0.00	0.39	0.05	0.68	20	0.55	0.36	0.01	0.08	0.52
21	0.47	0.52	0.00	0.01	0.63	22	0.46	0.54	0.00	0.00	0.36
23	0.25	0.75	0.00	0.00	0.04	24	0.22	0.78	0.00	0.00	0.00
25	0.21	0.01	0.78	0.00	0.32	26	0.21	0.63	0.00	0.16	0.03
27	0.18	0.11	0.11	0.60	0.22	28	0.18	0.72	0.00	0.10	0.00
29	0.15	0.15	0.44	0.26	0.18	30	0.07	0.93	0.00	0.00	0.00
31	0.04	0.55	0.20	0.21	0.00	32	0.03	0.97	0.00	0.00	0.00
33	0.00	0.00	1.00	0.00	0.10	34	0.00	1.00	0.00	0.00	0.00
35	0.00	0.00	0.92	0.08	0.10	36	0.00	0.94	0.00	0.06	0.00
37	0.03	0.01	0.96	0.00	0.13	38	0.02	0.29	0.00	0.69	0.00
39	0.01	0.97	0.00	0.02	0.00	40	0.01	0.29	0.02	0.68	0.00
41	0.00	0.24	0.00	0.76	0.00	42	0.00	0.93	0.00	0.07	0.00
43	0.00	1.00	0.00	0.00	0.00	44	0.00	1.00	0.00	0.00	0.00
45	0.00	0.98	0.02	0.00	0.00	46	0.00	0.00	0.00	1.00	0.06
47	0.00	1.00	0.00	0.00	0.00	48	0.00	0.96	0.00	0.04	0.00
49	0.00	0.91	0.00	0.09	0.00	50	0.00	0.76	0.03	0.21	0.00
51	0.00	0.00	0.32	0.68	0.07	52	0.00	0.13	0.02	0.85	0.01
53	0.00	0.82	0.15	0.03	0.00						

Table 4.3. The average expenditure of Group A and B.

Group	I	II	III	IV
A	0.8806	0.0514	0.0303	0.0377
B	0.1311	0.5277	0.1497	0.1915

hyperplane H_β. The hyperplane H_β is parallel to the two hyperplanes H_1 and H_2 such that it has a perpendicular distance of β from H_2.

The decision rule is applied to the remaining 22 captured customers. Among them, 6 customers (the first six customers in Table 4.4) belong to Group A and 12 customers belong to Group B. Their α values are computed and listed in Table 4.4. It is clear that if the value of β is set to be 0.75, all the customers will be classified correctly.

Table 4.4. The remaining one-third of the data for the validation of the HMM.

Customer	I	II	III	IV	α	Customer	I	II	III	IV	α
1'	0.98	0.00	0.02	0.00	1.00	2'	0.88	0.01	0.01	0.10	1.00
3'	0.74	0.26	0.00	0.00	0.76	4'	0.99	0.01	0.00	0.00	1.00
5'	0.99	0.01	0.00	0.00	1.00	6'	0.89	0.10	0.01	0.00	1.00
7'	0.00	0.00	1.00	0.00	0.10	8'	0.04	0.11	0.68	0.17	0.08
9'	0.00	0.02	0.98	0.00	0.09	10'	0.18	0.01	0.81	0.00	0.28
11'	0.32	0.05	0.61	0.02	0.41	12'	0.00	0.00	0.97	0.03	0.10
13'	0.12	0.14	0.72	0.02	0.16	14'	0.00	0.13	0.66	0.21	0.03
15'	0.00	0.00	0.98	0.02	0.10	16'	0.39	0.00	0.58	0.03	0.50
17'	0.27	0.00	0.73	0.00	0.38	18'	0.00	0.80	0.07	0.13	0.00

4.6 Summary

In this chapter, we propose a simple HMM with estimation methods. The framework of the HMM is simple and the model parameters can be estimated efficiently. Application to customers classification with practical data taken from a computer service company is presented and analyzed. Further disussions on new HMMs and applications will be given in Chapter 8.

5

Markov Decision Process for Customer Lifetime Value

5.1 Introduction

In this chapter a stochastic dynamic programming model with Markov chain is proposed to capture the customer behavior. The advantage of using the Markov chain is that the model can take into the account of the switch of the customers between the company and its competitors. Therefore customer relationships can be described in a probabilistic way, see for instance Pfeifer and Carraway [169]. Stochastic dynamic programming is then applied to solve the optimal allocation of promotion budget for maximizing the CLV. The proposed model is then applied to the practical data in a computer services company.

The customer equity should be measured in making the promotion plan so as to achieve an acceptable and reasonable budget. A popular approach is the Customer Lifetime Value (CLV). Kotler and Armstrong [134] defined a profitable customer as "a person, household, or company whose revenues over time exceeds, by an acceptable amount, the company costs consist of attracting, selling, and servicing that customer." This excess is called the CLV. In some literatures, CLV is also referred to "customer equity" [19]. In fact, some researchers define CLV as the customer equity less the acquisition cost. Nevertheless, in this thesis CLV is defined as the present value of the projected net cash flows that a firm expects to receive from the customer over time [42]. Recognizing the importance in decision making, CLV has been successfully applied in the problems of pricing strategy [18], media selection [115] and setting optimal promotion budget [22].

To calculate the CLV, a company should estimate the expected net cash flows receiving from the customer over time. The CLV is the present value of that stream of cash flows. However, it is a difficult task to estimate the net cash flows to be received from the customer. In fact, one needs to answer, for example, the following questions:

(i) How many customers one can attract given a specific advertising budget?
(ii) What is the probability that the customer will stay with the company?
(iii) How does this probability change with respect to the promotion budget?

To answer the first question, there are a number of advertising models, one can find in the book by Lilien, Kotler and Moorthy [146]. The second and the third questions give rise to an important concept, the retention rate. The retention rate [118] is defined as "the chance that the account will remain with the vendor for the next purchase, provided that the customer has bought from the vendor on each previous purchase". Jackson [118] proposed an estimation method for the retention rate based on historical data. Other models for the retention rate can also be found in [89, 146].

Blattberg and Deighton [22] proposed a formula for the calculation of CLV (customer equity). The model is simple and deterministic. Using their notations (see also [18, 19]), the CLV is the sum of two net present values: the return from acquisition spending and the return from retention spending. In their model, CLV is defined as

$$
\text{CLV} = \underbrace{am - A}_{acquisition} + \underbrace{\sum_{k=1}^{\infty} a(m - \frac{R}{r})[r(1+d)^{-1}]^k}_{retention}
$$

$$
= am - A + a(m - \frac{R}{r}) \times \frac{r}{(1+d-r)}
$$

(5.1)

where a is the acquisition rate, A is the level of acquisition spending, m is the margin on a transaction, R is the retention spending per customer per year, r is the yearly retention rate (a proportion) and d is the yearly discount rate appropriate for marketing investment. Moreover, they also assume that the acquisition rate a and retention rate r are functions of A and R respectively, and are given by

$$
a(A) = a_0(1 - e^{-K_1 A})
$$

and

$$
(R) = r_0(1 - e^{-K_2 R})
$$

where a_0 and r_0 are the estimated ceiling rates, K_1 and K_2 are two positive constants. In this chapter, a stochastic model (Markov decision process) is proposed for the calculation of CLV and the promotion planning.

The rest of the chapter is organized as follows. In Section 5.2, the Markov chain model for modelling the behavior of the customers is presented. In Section 5.3, stochastic dynamic programming is then used to calculate the CLV of the customers for three different scenarios:

(i) infinite horizon without constraint (without limit in the number of promotions),
(ii) finite horizon (with limited number of promotions), and
(iii) infinite horizon with constraints (with limited number of promotions).

In Section 5.4, we consider higher-order Markov decision process with applications to CLV problem. Finally a summary is given to conclude the chapter in Section 5.5.

5.2 Markov Chain Models for Customers' Behavior

In this section, Markov chain model for modelling the customers' behavior in a market is introduced. According to the usage of the customer, a company customer can be classified into N possible states

$$\{0, 1, 2, \ldots, N - 1\}.$$

Take for example, a customer can be classified into four states ($N = 4$): low-volume user (state 1), medium-volume user (state 2) and high-volume user (state 3) and in order to classify all customers in the market, state 0 is introduced. A customer is said to be in state 0, if he/she is either a customer of the competitor company or he/she did not purchase the service during the period of observation. Therefore at any time a customer in the market belongs to exactly one of the states in $\{0, 1, 2, \ldots, N - 1\}$. With these notations, a Markov chain is a good choice to model the transitions of customers among the states in the market.

A Markov chain model is characterized by an $N \times N$ transition probability matrix P. Here $P_{ij}(i, j = 0, 1, 2, \ldots, N - 1)$ is the transition probability that a customer will move to state i in the next period given that currently he/she is in state j. Hence the retention probability of a customer in state i ($i = 0, 1, \ldots, N - 1$) is given by P_{ii}. If the underlying Markov chain is assumed to be irreducible then the stationary distribution $\mathbf{p}$ exists, see for instance [180]. This means that there is an unique

$$\mathbf{p} = (p_0, p_1, \ldots, p_{N-1})^T$$

such that

$$\mathbf{p} = P\mathbf{p}, \quad \sum_{i=0}^{N-1} p_i = 1, \quad p_i \geq 0. \tag{5.2}$$

By making use of the stationary distribution $\mathbf{p}$, one can compute the retention probability of a customer as follows:

$$\sum_{i=1}^{N-1} \left(\frac{p_i}{\sum_{j=1}^{N-1} p_j} \right) (1 - P_{i0}) = 1 - \frac{1}{1 - p_0} \sum_{i=1}^{N-1} p_i P_{0i}$$

$$= 1 - \frac{p_0(1 - P_{00})}{1 - p_0}. \tag{5.3}$$

This is the probability that a customer will purchase service with the company in the next period. Apart from the retention probability, the Markov model can also help us in computing the CLV. In this case c_i is defined to be the revenue obtained from a customer in state i. Then the expected revenue is given by

$$\sum_{i=0}^{N-1} c_i p_i. \tag{5.4}$$

The above retention probability and the expected revenue are computed under the assumption that the company makes no promotion (in a non-competitive environment) through out the period. The transition probability matrix P can be significantly different when there is promotion making by the company. To demonstrate this, an application is given in the following subsection. Moreover, when promotions are allowed, what is the best promotion strategy such that the expected revenue is maximized? Similarly, what is the best strategy when there is a fixed budget for the promotions, e.g. the number of promotions is fixed? These issues will be discussed in the following section by using the stochastic dynamic programming model.

5.2.1 Estimation of the Transition Probabilities

In order to apply the Markov chain model, one has to estimate the transition probabilities from the practical data. In this subsection, an example in the computer service company is used to demonstrate the estimation. In the captured database of customers, each customer has four important attributes (A, B, C, D). Here A is the "Customer Number", each customer has an unique identity number. B is the "Week", the time (week) when the data was captured. C is the "Revenue" which is the total amount of money the customer spent in the captured week. D is the "Hour", the number of hours that the customer consumed in the captured week.

The total number of weeks of data available is 20. Among these 20 weeks, the company has a promotion for 8 consecutive weeks and no promotion for other 12 consecutive weeks. The behavior of customers in the period of promotion and no-promotion will be investigated. For each week, all the customers are classified into four states $(0, 1, 2, 3)$ according to the amount of "hours" consumed, see Table 5.1. We recall that a customer is said to be in state 0, if he/she is a customer of competitor company or he/she did not use the service for the whole week.

Table 5.1. The four classes of customers.

State	0	1	2	3
Hours	0.00	$1-20$	$21-40$	>40

From the data, one can estimate two transition probability matrices, one for the promotion period (8 consecutive weeks) and the other one for the no-promotion period (12 consecutive weeks). For each period, the number of customers switching from state i to state j is recorded. Then, divide it by the total number of customers in the state i, one can get the estimations for the one-step transition probabilities. Hence the transition probability matrices under the promotion period $P^{(1)}$ and the no-promotion period $P^{(2)}$ are given respectively below:

$$P^{(1)} = \begin{pmatrix} 0.8054 & 0.4163 & 0.2285 & 0.1372 \\ 0.1489 & 0.4230 & 0.3458 & 0.2147 \\ 0.0266 & 0.0992 & 0.2109 & 0.2034 \\ 0.0191 & 0.0615 & 0.2148 & 0.4447 \end{pmatrix}$$

and

$$P^{(2)} = \begin{pmatrix} 0.8762 & 0.4964 & 0.3261 & 0.2380 \\ 0.1064 & 0.4146 & 0.3837 & 0.2742 \\ 0.0121 & 0.0623 & 0.1744 & 0.2079 \\ 0.0053 & 0.0267 & 0.1158 & 0.2809 \end{pmatrix}.$$

$P^{(1)}$ is very different from $P^{(2)}$. In fact, there can be more than one type of promotion in general, the transition probability matrices for modelling the behavior of the customers can be more than two.

5.2.2 Retention Probability and CLV

The stationary distributions of the two Markov chains having transition probability matrices $P^{(1)}$ and $P^{(2)}$ are given respectively by

$$\mathbf{p}^{(1)} = (0.2306, 0.0691, 0.0738, 0.6265)^T$$

and

$$\mathbf{p}^{(2)} = (0.1692, 0.0285, 0.0167, 0.7856)^T.$$

The retention probabilities (cf. (5.3)) in the promotion period and no-promotion period are given respectively by 0.6736 and 0.5461. It is clear that the retention probability is significantly higher when the promotion is carried out.

From the customer data in the database, the average revenue of a customer is obtained in different states in both the promotion period and no-promotion period, see Table 5.2 below. We remark that in the promotion period, a big discount was given to the customers and therefore the revenue was significantly less than the revenue in the no-promotion period.

From (5.4), the expected revenue of a customer in the promotion period (assume that the only promotion cost is the discount rate) and no-promotion period are given by 2.42 and 17.09 respectively.

Although one can obtain the CLVs of the customers in the promotion period and the no-promotion period, one would expect to calculate the CLV in a mixture of promotion and no-promotion periods. Especially when the promotion budget is limited (the number of promotions is fixed) and one would like

Table 5.2. The average revenue of the four classes of customers.

State	0	1	2	3
Promotion	0.00	6.97	18.09	43.75
No-promotion	0.00	14.03	51.72	139.20

to obtain the optimal promotion strategy. Stochastic dynamic programming with Markov process provides a good approach for solving the above problems. Moreover, the optimal stationary strategy for the customers in different states can also be obtained by solving the stochastic dynamic programming problem.

5.3 Stochastic Dynamic Programming Models

The problem of solving the optimal promotion strategy can be fitted into the framework of stochastic dynamic programming models. In this section, stochastic dynamic programming models are presented for maximizing the CLV under optimal promotion strategy. The notations of the model are given as follows:

(i) N, the total number of states (indexed by $i = 0, 1, \ldots, N - 1$);

(ii) A_i, the set containing all the actions in state i (indexed by k);

(iii) T, number of months remained in the planning horizon (indexed by $t = 1, \ldots, T$);

(iv) d_k, the resources required for carrying out the action k in each period;

(v) $c_i^{(k)}$, the revenue obtained from a customer in state i with the action k in each period;

(vi) $p_{ij}^{(k)}$, the transition probability for customer moving from state j to state i under the action k in each period;

(vii) α, discount rate.

Similar to the MDP introduced in Chapter 1, the value of an optimal policy $v_i(t)$ is defined to be the total expected revenue obtained in the stochastic dynamic programming model with t months remained for a customer in state i for $i = 0, 1, \ldots, N - 1$ and $t = 1, 2, \ldots, T$. Therefore, the recursive relation for maximizing the revenue is given as follows:

$$v_i(t) = \max_{k \in A_i} \left\{ c_i^{(k)} - d_k + \alpha \sum_{j=0}^{N-1} p_{ji}^{(k)} v_j(t-1) \right\}. \tag{5.5}$$

In the following subsections, three different CLV models based on the above recursive relation are considered. They are infinite horizon without constraints, finite horizon with hard constraints and infinite horizon with constraints. For each case, an application with practical data in a computer service company is presented.

5.3.1 Infinite Horizon without Constraints

The problem is considered as an infinite horizon stochastic dynamic programming. From the standard results in stochastic dynamic programming [209], for each i, the optimal values v_i for the discounted infinite horizon Markov decision process satisfy the relationship

$$v_i = \max_{k \in A_i} \left\{ c_i^{(k)} - d_k + \alpha \sum_{j=0}^{N-1} p_{ji}^{(k)} v_j \right\}. \tag{5.6}$$

Therefore we have

$$v_i \geq c_i^{(k)} - d_k + \alpha \sum_{j=0}^{N-1} p_{ji}^{(k)} v_j \tag{5.7}$$

for each i. In fact, the optimal values v_i are the smallest numbers (the least upper bound over all possible policy values) that satisfy these inequalities. This suggests that the problem of determining the v_i's can be transformed into the following linear programming problem [4, 208, 209]:

$$\left\{ \begin{array}{l} \min \quad x_0 = \sum_{i=0}^{N-1} v_i \\ \text{subject to} \\ \quad v_i \geq c_i^{(k)} - d_k + \alpha \sum_{j=0}^{N-1} p_{ji}^{(k)} v_j, \quad \text{for} \quad i = 0, \ldots, N-1; \\ \quad v_i \geq 0 \quad \text{for} \quad i = 0, \ldots, N-1. \end{array} \right. \tag{5.8}$$

The above linear programming problem can be solved easily by using EXCEL spreadsheet. In addition, a demonstration EXCEL file is available at the following site [224], see also Fig 5.1 (Taken from [70]). Return to the model for the computer service company, there are 2 actions available (either (P) promotion or (NP) no-promotion) for all possible states. Thus $A_i = \{P, NP\}$ for all $i = 0, \ldots, N-1$. Moreover, customers are classified into 4 clusters, therefore $N = 4$ (possible states of a customer are $0, 1, 2, 3$). Since no promotion cost is incurred for the action (NP), therefore $d_{NP} = 0$. For simplification, d is used to denote the only promotion cost instead of d_P in the application.

Table 5.4 presents optimal stationary policies (i.e., to have promotion $D_i = P$ or no-promotion $D_i = NP$ depends on the state i of customer) and the corresponding revenues for different discount factors α and fixed promotion

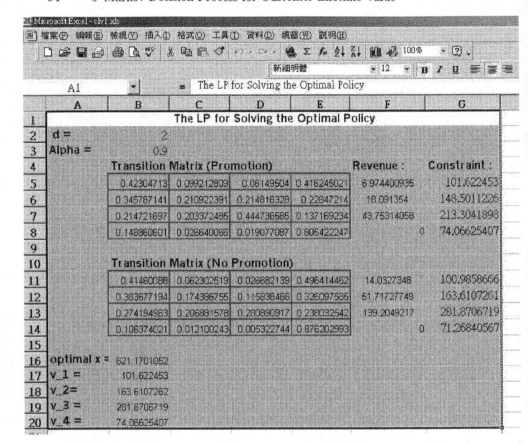

Fig. 5.1. EXCEL for solving infinite horizon problem without constraint.

costs d. For instance, when the promotion cost is 0 and the discount factor is 0.99, then the optimal strategy is that when the current state is 0 or 1, the promotion should be done i.e. $D_0 = D_1 = P$, and when the current state is 2 or 3, no promotion is required, i.e. $D_2 = D_3 = NP$, (see the first column of the upper left hand box of Table 5.3). The other values can be interpreted similarly. From the numerical examples, the following conclusions are drawn.

- When the fixed promotion cost d is large, the optimal strategy is that the company should not conduct any promotion on the active customers and should only conduct promotion scheme to both inactive (purchase no service) customers and customers of the competitor company. However, when d is small, the company should take care of the low-volume customers to prevent this group of customers from churning to the competitor companies.
- It is also clear that the CLV of a high-volume user is larger than the CLV of other groups.

- The CLVs of each group depend on the discount rate α significantly. Here the discount rate can be viewed as the technology depreciation of the computer services in the company. Therefore, in order to generate the revenue of the company, new technology and services should be provided.

Table 5.3. Optimal stationary policies and their CLVs.

| | $d = 0$ | | | $d = 1$ | | | $d = 2$ | | |
	$\alpha = 0.99$	$\alpha = 0.95$	$\alpha = 0.90$	$\alpha = 0.99$	$\alpha = 0.95$	$\alpha = 0.90$	$\alpha = 0.99$	$\alpha = 0.95$	$\alpha = 0.90$
x_0	4791	1149	687	4437	1080	654	4083	1012	621
v_0	1112	204	92	1023	186	83	934	168	74
v_1	1144	234	119	1054	216	110	965	198	101
v_2	1206	295	179	1118	278	171	1030	261	163
v_3	1328	415	296	1240	399	289	1153	382	281
D_0	P	P	P	P	P	P	P	P	P
D_1	P	P	P	P	P	P	P	P	P
D_2	NP	NP	NP	NP	NP	NP	NP	NP	NP
D_3	NP	NP	NP	NP	NP	NP	NP	NP	NP

| | $d = 3$ | | | $d = 4$ | | | $d = 5$ | | |
	$\alpha = 0.99$	$\alpha = 0.95$	$\alpha = 0.90$	$\alpha = 0.99$	$\alpha = 0.95$	$\alpha = 0.90$	$\alpha = 0.99$	$\alpha = 0.95$	$\alpha = 0.90$
x_0	3729	943	590	3375	879	566	3056	827	541
v_0	845	151	65	755	134	58	675	119	51
v_1	877	181	94	788	164	88	707	151	82
v_2	942	245	156	854	230	151	775	217	145
v_3	1066	366	275	978	351	269	899	339	264
D_0	P	P	P	P	P	P	P	P	P
D_1	P	P	NP	P	NP	NP	NP	NP	NP
D_2	NP	NP	NP	NP	NP	NP	NP	NP	NP
D_3	NP	NP	NP	NP	NP	NP	NP	NP	NP

5.3.2 Finite Horizon with Hard Constraints

In the computer service and telecommunication industry, the product life cycle is short, e.g., it is usually one year. Therefore, the case of finite horizon with limited budget constraint is considered. This problem can also be solved efficiently by using stochastic dynamic programming and the optimal revenues obtained in the previous section is used as the boundary conditions.

The model's parameters are defined as follows:

$$
\begin{aligned}
n &= \text{number of weeks remaining;} \\
p &= \text{number of possible promotions remaining.}
\end{aligned}
$$

The recursive relation for the problem is given as follows:

$$
v_i(n, p) = \max \{ c_i^{(P)} - d_P + \alpha \textstyle\sum_{j=0}^{N-1} p_{ji}^{(P)} v_j(n-1, p-1), \\
c_i^{(NP)} - d_{NP} + \alpha \textstyle\sum_{j=0}^{N-1} p_{ji}^{(NP)} v_j(n-1, p) \} \tag{5.9}
$$

for $n = 1, \ldots, n_{max}$ and $p = 1, \ldots, p_{max}$ and

$$
v_i(n, 0) = c_i^{(NP)} - d_{NP} + \alpha \sum_{j=0}^{N-1} p_{ji}^{(NP)} v_j(n-1, 0) \tag{5.10}
$$

for $n = 1, \ldots, n_{max}$. The above dynamic programming problem can be solved easily by using spreadsheet EXCEL. A demonstration EXCEL file can be found at the following site [225], see also Fig. 5.2 (Taken from [70]). In the numerical experiment of the computer service company, the length of planning period is set to be $n_{max} = 52$ and the maximum number of promotions is $p_{max} = 4$. By solving the dynamic programming problem, the optimal values and promotion strategies are listed in Table 5.4. The optimal solution in the table is presented as follows:

$$
(t_1, t_2, t_3, t_4, r^*),
$$

where r^* is the optimal expected revenue, and t_i is the promotion week of the optimal promotion strategy and "-" means no promotion. Findings are summarized as follows:

- For different values of the fixed promotion cost d, the optimal strategy for the customers in states 2 and 3 is to conduct no promotion.
- While for those in state 0, the optimal strategy is to conduct all the four promotions as early as possible.
- In state 1, the optimal strategy depends on the value of d. If d is large, then no promotion will be conducted. However, when d is small, promotions are carried out and the strategy is to put the promotions as late as possible.

5.3.3 Infinite Horizon with Constraints

For comparisons, the model in Section 5.3.2 is extended to the infinite horizon case. Similar to the previous model, the finite number of promotions available is denoted by p_{max}. Then the value function $v_i(p)$, which represents the optimal discounted utility starting at state i when there are p number of promotions remaining, is the unique fixed point of the equations:

Fig. 5.2. EXCEL for solving finite horizon problem without constraint.

$$v_i(p)$$
$$= \max \left\{ c_i^{(P)} - d_P + \alpha \sum_{j=0}^{N-1} p_{ji}^{(P)} v_j(p-1), c_i^{(NP)} - d_{NP} + \alpha \sum_{j=0}^{N-1} p_{ji}^{(NP)} v_j(p) \right\},$$
(5.11)

for $p = 1, \ldots, p_{\max}$, and

$$v_i(0) = c_i^{(NP)} - d_{NP} + \alpha \sum_{j=0}^{N-1} p_{ji}^{(NP)} v_j(0).$$
(5.12)

Table 5.4. Optimal promotion strategies and their CLVs.

α	State 0	State 1	State 2	State 3
0.9	$(1,2,3,4,67)$	$(1,45,50,52,95)$	$(-,-,-,-,158)$	$(-,-,-,-,276)$
$d=0$ 0.95	$(1,2,3,4,138)$	$(45,48,50,51,169)$	$(-,-,-,-,234)$	$(-,-,-,-,335)$
0.99	$(1,2,3,4,929)$	$(47,49,50,51,963)$	$(-,-,-,-,1031)$	$(-,-,-,-,1155)$
0.9	$(1,2,3,4,64)$	$(47,49,51,52,92)$	$(-,-,-,-,155)$	$(-,-,-,-,274)$
$d=1$ 0.95	$(1,2,3,4,133)$	$(47,49,51,52,164)$	$(-,-,-,-,230)$	$(-,-,-,-,351)$
0.99	$(1,2,3,4,872)$	$(47,49,51,52,906)$	$(-,-,-,-,974)$	$(-,-,-,-,1098)$
0.9	$(1,2,3,4,60)$	$(49,50,51,52,89)$	$(-,-,-,-,152)$	$(-,-,-,-,271)$
$d=2$ 0.95	$(1,2,3,4,128)$	$(48,50,51,52,160)$	$(-,-,-,-,225)$	$(-,-,-,-,347)$
0.99	$(1,2,3,4,815)$	$(48,49,51,52,849)$	$(-,-,-,-,917)$	$(-,-,-,-,1041)$
0.9	$(1,2,3,4,60)$	$(-,-,-,-,87)$	$(-,-,-,-,150)$	$(-,-,-,-,269)$
$d=3$ 0.95	$(1,2,3,4,123)$	$(49,50,51,52,155)$	$(-,-,-,-,221)$	$(-,-,-,-,342)$
0.99	$(1,2,3,4,758)$	$(48,50,51,52,792)$	$(-,-,-,-,860)$	$(-,-,-,-,984)$
0.9	$(1,2,3,4,54)$	$(-,-,-,-,84)$	$(-,-,-,-,147)$	$(-,-,-,-,266)$
$d=4$ 0.95	$(1,2,3,4,119)$	$(-,-,-,-,151)$	$(-,-,-,-,217)$	$(-,-,-,-,338)$
0.99	$(1,2,3,4,701)$	$(49,50,51,52,736)$	$(-,-,-,-,804)$	$(-,-,-,-,928)$
0.9	$(1,2,3,4,50)$	$(-,-,-,-,81)$	$(-,-,-,-,144)$	$(-,-,-,-,264)$
$d=5$ 0.95	$(1,2,3,4,114)$	$(-,-,-,-,147)$	$(-,-,-,-,212)$	$(-,-,-,-,334)$
0.99	$(1,2,3,4,650)$	$(-,-,-,-,684)$	$(-,-,-,-,752)$	$(-,-,-,-,876)$

Since $[p_{ij}^{(k)}]$ is a transition probability matrix, the set of linear equations (5.12) with four unknowns has a unique solution. We note that (5.11) can be computed by the value iteration algorithm, i.e. as the limit of $v_i(n,p)$ (computed in Section 5.3.2), as n tends to infinity. Alternatively, it can be solved by linear programming [4]:

$$
\begin{cases}
\min \quad x_0 = \displaystyle\sum_{i=0}^{N-1} \sum_{p=1}^{p_{\max}} v_i(p) \\[2mm]
\text{subject to} \\[1mm]
v_i(p) \geq c_i^{(P)} - d_1 + \alpha \displaystyle\sum_{j=0}^{N-1} p_{ji}^{(P)} v_j(p-1), \\[1mm]
\text{for} \quad i = 0, \ldots, N-1, p = 1, \ldots, p_{\max}; \\[2mm]
v_i(p) \geq c_i^{(NP)} - d_2 + \alpha \displaystyle\sum_{j=0}^{N-1} p_{ji}^{(NP)} v_j(p), \\[1mm]
\text{for} \quad i = 0, \ldots, N-1, p = 1, \ldots, p_{\max}.
\end{cases}
$$

We note that $v_i(0)$ is not included in the linear programming constraints and the objective function; $v_i(0)$ is solved before hand using (5.12). A demonstration EXCEL file can be found at the following site [226], see also Fig. 5.3 (Taken from [70]).

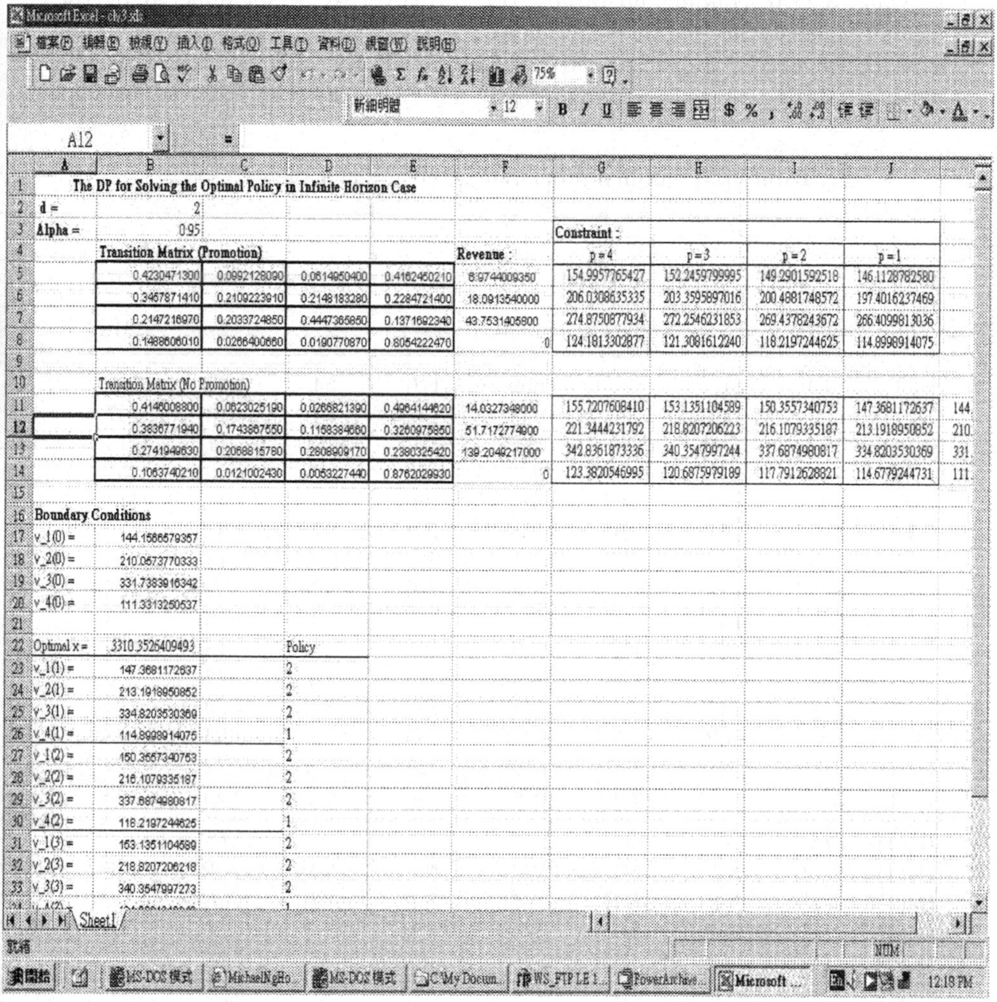

Fig. 5.3. EXCEL for solving infinite horizon problem with constraints.

Tables 5.5 and 5.6 give the optimal values and promotion strategies of the computer service company. For instance, when the promotion cost is 0 and the discount factor is 0.99, then the optimal strategy is that when the current state is 1, 2 or 3, the promotion should be done when there are some available promotions, i.e. $D_1(p) = D_2(p) = D_3(p) = P$ for $p = 1, 2, 3, 4$, and when the current state is 0, no promotion is required, i.e. $D_0(p) = NP$ for $p = 1, 2, 3, 4$. Their corresponding CLVs $v_i(p)$ for different states and different numbers of remaining promotion are also listed (see the first column in the left hand side of Table 5.6.

From Tables 5.5 and 5.6, the optimal strategy for the customers in states 1, 2 and 3 is to conduct no promotion. Moreover, it is not affected by the promotion cost and the discount factor. These results are slightly different from those for the finite horizon case. However, the optimal strategy is to conduct all the four promotions to customer with state 0 as early as possible.

Table 5.5. Optimal promotion strategies and their CLVs.

	$d = 0$			$d = 1$			$d = 2$		
	$\alpha = 0.99$	$\alpha = 0.95$	$\alpha = 0.90$	$\alpha = 0.99$	$\alpha = 0.95$	$\alpha = 0.90$	$\alpha = 0.99$	$\alpha = 0.95$	$\alpha = 0.9$
x_0	11355	3378	2306	11320	3344	2277	11277	3310	2248
$v_0(1)$	610	117	55	609	116	54	608	115	53
$v_1(1)$	645	149	85	644	148	84	643	147	84
$v_2(1)$	713	215	149	712	214	148	711	213	147
$v_3(1)$	837	337	267	836	336	267	845	335	266
$v_0(2)$	616	122	60	614	120	58	612	118	56
$v_1(2)$	650	154	89	648	152	87	647	150	86
$v_2(2)$	718	219	152	716	218	151	714	216	149
$v_3(2)$	842	341	271	840	339	269	839	338	268
$v_1(3)$	656	158	92	654	156	90	650	153	88
$v_2(3)$	724	224	155	722	221	153	718	219	151
$v_3(3)$	848	345	273	846	343	271	842	340	270
$v_0(4)$	628	131	67	624	128	63	620	124	60
$v_1(4)$	662	162	95	658	159	92	654	158	89
$v_2(4)$	730	228	157	726	225	155	722	221	152
$v_3(4)$	854	349	276	850	346	273	846	343	271
$D_0(1)$	P	P	P	P	P	P	P	P	P
$D_1(1)$	NP	NP	NP	NP	NP	NP	NP	NP	NP
$D_2(1)$	NP	NP	NP	NP	NP	NP	NP	NP	NP
$D_3(1)$	NP	NP	NP	NP	NP	NP	NP	NP	NP
$D_0(2)$	P	P	P	P	P	P	P	P	P
$D_1(2)$	NP	NP	NP	NP	NP	NP	NP	NP	NP
$D_2(2)$	NP	NP	NP	NP	NP	NP	NP	NP	NP
$D_3(2)$	NP	NP	NP	NP	NP	NP	NP	NP	NP
$D_0(3)$	P	P	P	P	P	P	P	P	P
$D_1(3)$	NP	NP	NP	NP	NP	NP	NP	NP	NP
$D_3(3)$	NP	NP	NP	NP	NP	NP	NP	NP	NP
$D_0(4)$	P	P	P	P	P	P	P	P	P
$D_1(4)$	NP	NP	NP	NP	NP	NP	NP	NP	NP
$D_2(4)$	NP	NP	NP	NP	NP	NP	NP	NP	NP
$D_3(4)$	NP	NP	NP	NP	NP	NP	NP	NP	NP

Table 5.6. Optimal promotion strategies and their CLVs.

	$d = 3$			$d = 4$			$d = 5$		
	$\alpha = 0.99$	$\alpha = 0.95$	$\alpha = 0.90$	$\alpha = 0.99$	$\alpha = 0.95$	$\alpha = 0.90$	$\alpha = 0.99$	$\alpha = 0.95$	$\alpha = 0.90$
x_0	11239	3276	2218	11200	3242	2189	11161	3208	2163
$v_0(1)$	607	114	52	606	113	51	605	112	50
$v_1(1)$	641	146	83	641	146	82	640	145	81
$v_2(1)$	710	212	146	709	211	145	708	211	145
$v_3(1)$	834	334	265	833	333	264	832	332	264
$v_0(2)$	610	116	54	608	114	52	606	112	50
$v_1(2)$	645	149	84	643	147	83	641	145	81
$v_2(2)$	713	214	148	711	213	146	709	211	145
$v_3(2)$	837	336	266	835	334	265	833	333	264
$v_0(3)$	613	119	56	610	116	53	607	113	50
$v_1(3)$	647	151	86	645	148	83	642	146	81
$v_2(3)$	715	216	149	713	214	147	710	211	145
$v_3(3)$	839	338	268	837	336	266	834	333	264
$v_0(4)$	616	121	57	612	117	54	608	113	50
$v_1(4)$	650	152	87	646	149	84	643	146	81
$v_2(4)$	718	218	150	714	215	147	711	212	145
$v_3(4)$	842	340	269	838	337	266	835	334	265
$D_0(1)$	P	P	P	P	P	P	P	P	P
$D_1(1)$	NP	NP	NP	NP	NP	NP	NP	NP	NP
$D_2(1)$	NP	NP	NP	NP	NP	NP	NP	NP	NP
$D_3(1)$	NP	NP	NP	NP	NP	NP	NP	NP	NP
$D_0(2)$	P	P	P	P	P	P	P	P	P
$D_1(2)$	NP	NP	NP	NP	NP	NP	NP	NP	NP
$D_2(2)$	NP	NP	NP	NP	NP	NP	NP	NP	NP
$D_3(2)$	NP	NP	NP	NP	NP	NP	NP	NP	NP
$D_0(3)$	P	P	P	P	P	P	P	P	P
$D_1(3)$	NP	NP	NP	NP	NP	NP	NP	NP	NP
$D_2(3)$	NP	NP	NP	NP	NP	NP	NP	NP	NP
$D_3(3)$	NP	NP	NP	NP	NP	NP	NP	NP	NP
$D_0(4)$	P	P	P	P	P	P	P	P	P
$D_1(4)$	NP	NP	NP	NP	NP	NP	NP	NP	NP
$D_2(4)$	NP	NP	NP	NP	NP	NP	NP	NP	NP
$D_3(4)$	NP	NP	NP	NP	NP	NP	NP	NP	NP

5.4 Higher-order Markov decision process

The MDP presented in previous section is a first-order type, i.e., the transition probabilities depend on the current state only. A brief introduction has been given in Chapter 1. For the HDMP, the transition probabilities depend on the current state and a number of previous states. For instance, the probabilities of a second-order MDP moving from state s_i to state s_j depend only on the latest two states, the present state s_i and the previous state s_h. The transition probability is denoted by p_{hij}. In this section, we are interested in studying a Higher-order Markov Decision Process (HMDP) with applications to the CLV problems.

In the infinite horizon case, there are infinite number of policies with the initial state s_i and the previous state s_h. The policy D prescribes an alternative, say k^*, for the transition out of states s_h and state s_i. The probability of being in state s_j after one transition is $p_{hij}^{(k*)}$ and this probability is rewritten as $p(1, j)$. Now using the alternatives directed by D, one can calculate the probabilities of being in the various states after two transitions; these probabilities can be denoted by

$$p(2, l) \quad \text{for} \quad l = 0, 1, \ldots, N - 1.$$

Similarly one can calculate the probability $p(n, j)$ of being in state s_i and state s_h after n transitions. Denoting by $D(n, h, i)$ the alternative that D prescribes for use after n transitions if the system is in state s_j, the expected reward to be earned by D on the $(n + 1)$-transition would be

$$\sum_{j=0}^{N-1} p(n, j) q_j^{D(n,h,i)} \tag{5.13}$$

and the present value of this sum is

$$\alpha^n \sum_{j=0}^{N-1} p(n, j) q_j^{D(n,h,i)}. \tag{5.14}$$

Thus the total expected reward of D is given by

$$q_i^{(k*)} + \sum_{n=1}^{\infty} \alpha^n \sum_{j=0}^{N-1} p(n, j) q_j^{D(n,h,i)}. \tag{5.15}$$

Choosing Q such that

$$|q_l^{(k)}| \leq Q \quad \text{for all} \quad l = 0, 1, \ldots, N - 1. \tag{5.16}$$

and $k \in A_i$, the sum is absolutely convergent. This sum is called the value of the policy D, and it is denoted by $w_{hi}(D)$. It is clear that

$$|w_{hi}(D)| \leq Q(1 - \alpha)^{-1}. \tag{5.17}$$

5.4.1 Stationary policy

A stationary policy is a policy that the choice of alternative depends only on the state the system is in and is independent of n. $D(h, i)$ is defined to be the stationary policy with the current state s_i and the previous s_h. For a Markov decision process with infinite horizon and discount factor α, $0 < \alpha < 1$, the value of an optimal policy is defined as follows:

$$v_{hi} = \text{lub }\{w_{hi}(D)|D \text{ a policy with initial state } s_i \text{ and previous state } s_h\} \quad (5.18)$$

where lub is the standard abbreviation for least upper bound.

Proposition 5.1. *For a Markov decision process with infinite horizon, discount factor α, where $0 < \alpha < 1$, and*

$$u_{hi} = \max_{k \in A_i}\{q_i^{(k)} + \alpha \sum_{j=0}^{N-1} p_{hij}^{(k)} v_{ij}\}, \quad h, i = 0, 1, \ldots, N-1. \quad (5.19)$$

Then, for each h, i, $u_{hi} = v_{hi}$.

Proof. Fixing $h, i = 0, 1, \ldots, N-1$, let D be any policy with initial state s_i and previous state s_h. Suppose D prescribes alternative k^* on the first transition out of s_h, s_i; and denote by $\bar{D}_{ij}$ the associated one-step-removed policy. Then

$$w_{hi}(D) = q_i^{(k^*)} + \alpha \sum_{j=0}^{N-1} p_{hij}^{(k^*)} w_{ij}(\bar{D}_{ij})$$

$$\leq q_i^{(k^*)} + \alpha \sum_{j=0}^{N-1} p_{hij}^{(k^*)} v_{ij}$$

$$\leq \max_{k \in A_i}\{q_i^{(k)} + \alpha \sum_{j=0}^{N-1} p_{hij}^{(k)} v_{ij}\} = u_{hi}.$$

Therefore u_{hi} is an upper bound for the set

$$\{w_{hi}(D)|D \text{ a policy with initial state } s_i \text{ previous state } s_h\}$$

and

$$v_{hi} = \text{lub }\{w_{hi}(D)\} \leq u_{hi}.$$

Considering an alternative k_{hi} such that

$$u_{hi} = \max_{k \in A_i}\{q_i^{(k)} + \alpha \sum_{j=0}^{N-1} p_{hij}^{(k)} v_{ij}\} = q_i^{(k_{hi})} + \alpha \sum_{j=0}^{N-1} p_{hij}^{(k_{hi})} v_{ij}.$$

For any given $\epsilon > 0$ and for each j, a policy D_{hi}^* is chosen with initial state s_i and previous state s_h such that

$$v_{hi} - \epsilon < w_{hi}(D_{hi}^*).$$

Define a policy D with initial state s_i and previous state s_h as follows: use alternative k_{hi} out of states s_h and state s_i, then for each h, i if the system moves to state s_j on the first transition, policy D_{ij}^* is used thereafter. We have

$$u_{hi} = q_i^{(k_{hi})} + \alpha \sum_{j=0}^{N-1} p_{hij}^{(k_{hi})} v_{ij}$$

$$\leq q_i^{(k_{hi})} + \alpha \sum_{j=0}^{N-1} p_{hij}^{(k_{hi})} (w_{ij}(D_{ij}^*) + \epsilon)$$

$$= q_i^{(k_{hi})} + \alpha \sum_{j=0}^{N-1} p_{hij}^{(k_{hi})} w_{ij}(D_{ij}^*) + \alpha\epsilon \sum_{j=0}^{N-1} p_{hij}^{(k_{hi})}$$

$$= w_{hi}(D) + \alpha\epsilon$$

$$< v_{hi} + \epsilon.$$

Since ϵ is arbitrary, $u_{hi} \leq v_{hi}$. The result follows.

Proposition 5.2. *(Stationary Policy Theorem) Given a Markov decision process with infinite horizon and discount factor α, $0 < \alpha < 1$, choose, for each h, i, an alternative k_{hi} such that*

$$\max_{k \in A_i}\{q_i^{(k)} + \alpha \sum_{j=0}^{N-1} p_{hij}^{(k)} v_{ij}\} = q_i^{(k_{hi})} + \alpha \sum_{j=0}^{N-1} p_{hij}^{(k_{hi})} v_{ij}.$$

Define the stationary policy D by $D(h, i) = k_{hi}$. Then for each h, i, $w_{hi}(D) = v_{hi}$.

Proof. Since

$$v_{hi} = q_i^{(k_{hi})} + \alpha \sum_{j=0}^{N-1} p_{hij}^{(k_{hi})} v_{ij},$$

we have

$$\mathbf{v} = \mathbf{q} + \alpha P \mathbf{v}$$

where

$$\mathbf{v} = [v_{0,0}, v_{0,1}, \ldots v_{0,N-1}, v_{1,0}, \ldots v_{N-1,N-1}]^T,$$

$$\mathbf{q} = [q_0, q_1, \ldots, q_{N-1}, q_0, \ldots, q_{N-1}]^T,$$

and

$$P = [p_{hij}^{(k_{hi})}].$$

The superscript are omitted in the above vectors. For $0 < \alpha < 1$, the matrix $(I - \alpha P)$ is nonsingular and the result follows.

According to the above two propositions, the optimal stationary policy can be obtained by solving the following LP problem:

$$
\begin{cases}
\min\left\{x_{0,0} + x_{0,1} + \cdots + x_{0,N-1} + x_{1,0} + \cdots + x_{N-1,N-1}\right\} \\
\text{subject to} \\
\qquad x_{hi} \geq q_i^{(k)} + \alpha \sum_{j=0}^{N-1} p_{hij}^{(k)} x_{ij}, \quad h, i = 0, 1, \ldots, N-1, \\
\qquad k \in A_i.
\end{cases}
\tag{5.20}
$$

5.4.2 Application to the calculation of CLV

In previous sections, a first-order MDP is applied to a computer service company. In this section, the same set of customers' database is used with the HMDP. Comparison of two models will be given, Ching et al. [72].

The one-step transition probabilities are given in Section 5.3. Similarly, one can estimate the second-order (two-step) transition probabilities. Given that the current state i and previous state h, the number of customers switching to state j is recorded. Then, divide it by the total number of customers in the current state i and previous state j. The values obtained are the second-order transition probabilities. The transition probabilities under the promotion and no-promotion period are given respectively in Table 5.7.

Table 5.7. The second-order transition probabilities.

States	Promotion				No-Promotion			
	0	1	2	3	0	1	2	3
(0,0)	0.8521	0.1225	0.0166	0.0088	0.8957	0.0904	0.0098	0.0041
(0,1)	0.5873	0.3258	0.0549	0.0320	0.6484	0.3051	0.0329	0.0136
(0,2)	0.4471	0.3033	0.1324	0.1172	0.5199	0.3069	0.0980	0.0753
(0,3)	0.3295	0.2919	0.1482	0.2304	0.4771	0.2298	0.1343	0.1587
(1,0)	0.6739	0.2662	0.0394	0.0205	0.7287	0.2400	0.0227	0.0086
(1,1)	0.3012	0.4952	0.1661	0.0375	0.3584	0.5117	0.1064	0.0234
(1,2)	0.1915	0.4353	0.2169	0.1563	0.2505	0.4763	0.1860	0.0872
(1,3)	0.1368	0.3158	0.2271	0.3203	0.1727	0.3750	0.2624	0.1900
(2,0)	0.5752	0.2371	0.1043	0.0834	0.6551	0.2253	0.0847	0.0349
(2,1)	0.2451	0.4323	0.2043	0.1183	0.3048	0.4783	0.1411	0.0757
(2,2)	0.1235	0.3757	0.2704	0.2304	0.2032	0.3992	0.2531	0.1445
(2,3)	0.1030	0.2500	0.2630	0.3840	0.1785	0.2928	0.2385	0.2901
(3,0)	0.4822	0.2189	0.1496	0.1494	0.6493	0.2114	0.0575	0.0818
(3,1)	0.2263	0.3343	0.2086	0.2308	0.2678	0.4392	0.1493	0.1437
(3,2)	0.1286	0.2562	0.2481	0.3671	0.2040	0.3224	0.2434	0.2302
(3,3)	0.0587	0.1399	0.1855	0.6159	0.1251	0.1968	0.1933	0.4848

The transition probability from state 0 to state 0 is very high in the first-order model for both promotion and no-promotion period. However, in the second-order model, the transition probabilities

$$(0,0) \to 0, (1,0) \to 0, (2,0) \to 0 \quad \text{and} \quad (3,0) \to 0$$

are very different. It is clear that the second-order Markov chain model can better capture the customers' behavior than the first-order Markov chain model.

In Tables 5.8, 5.9 and 5.10, the optimal stationary policy is given for the first-order and the second-order MDP respectively for different values of discount factor α and promotion cost d. Once again, (P) represents to conduct promotion and (NP) represents to make no promotion. It is found that the optimal stationary policies for both models are consistent in the sense that $D_i = D_{ii}$ for $i = 0, 1, 2, 3$ in all the tested cases. For the second-order case, the optimal stationary policy D_{ii} depends not only on states (the optimal policy depends on the current state only in the first-order model) but also on the value of α and d. It is observed that the second-order Markov decision process always gives better objective value.

5.5 Summary

Finally, we end this chapter by the following summary. In this chapter, stochastic dynamic programming models are proposed for the optimization of CLV. Both cases of infinite horizon and finite horizon with budget constraints are discussed. The former case can be solved by using linear programming techniques, the later problem can be solved by using dynamic programming approach. For both cases, they can be implemented easily in an EXCEL spreadsheet. The models are then applied to practical data of a computer service company. The company makes use of the proposed CLV model to make and maintain value-laden relationships with the customers. We also extend the idea of MDP to a higher-order setting. Optimal stationary policy is also obtained in this case.

Further research can be done in promotion strategy through advertising. Advertising is an important tool in modern marketing. The purpose of advertising is to enhance potential users' responses to the company by providing information for choosing a particular product or service. A number of marketing models can be found in Lilien et al. [146] and the references therein. It has been shown that a pulsation advertising policy is effective, Mesak et al. [150, 151, 152, 153] and Ching et al. [74]. It will be interesting to incorporate the pulsation advertising policy in the CLV model.

Table 5.8. Optimal strategies when the first-order MDP is used.

	$d = 0$			$d = 1$			$d = 2$		
	$\alpha = 0.99$	$\alpha = 0.95$	$\alpha = 0.90$	$\alpha = 0.99$	$\alpha = 0.95$	$\alpha = 0.90$	$\alpha = 0.99$	$\alpha = 0.95$	$\alpha = 0.90$
x_0	4791	1149	687	4437	1080	654	4083	1012	621
v_0	1112	204	92	1023	186	83	934	168	74
v_1	1144	234	119	1054	216	110	965	198	101
v_2	1206	295	179	1118	278	171	1030	261	163
v_3	1328	415	296	1240	399	289	1153	382	281
D_0	P	P	P	P	P	P	P	P	P
D_1	P	P	P	P	P	P	P	P	P
D_2	NP	NP	NP	NP	NP	NP	NP	NP	NP
D_3	NP	NP	NP	NP	NP	NP	NP	NP	NP

	$d = 3$			$d = 4$			$d = 5$		
	$\alpha = 0.99$	$\alpha = 0.95$	$\alpha = 0.90$	$\alpha = 0.99$	$\alpha = 0.95$	$\alpha = 0.90$	$\alpha = 0.99$	$\alpha = 0.95$	$\alpha = 0.90$
x_0	3729	943	590	3375	879	566	3056	827	541
v_0	845	151	65	755	134	58	675	119	51
v_1	877	181	94	788	164	88	707	151	82
v_2	942	245	156	854	230	151	775	217	145
v_3	1066	366	275	978	351	269	899	339	264
D_0	P	P	P	P	P	P	P	P	P
D_1	P	P	NP	P	NP	NP	NP	NP	NP
D_2	NP	NP	NP	NP	NP	NP	NP	NP	NP
D_3	NP	NP	NP	NP	NP	NP	NP	NP	NP

Table 5.9. Optimal strategies when the second-order MDP is used.

	$d = 0$			$d = 1$			$d = 2$		
	$\alpha = 0.99$	$\alpha = 0.95$	$\alpha = 0.90$	$\alpha = 0.99$	$\alpha = 0.95$	$\alpha = 0.90$	$\alpha = 0.99$	$\alpha = 0.95$	$\alpha = 0.90$
x_0	19001	5055	3187	17578	4785	3066	16154	4520	2950
v_{00}	1034	177	74	943	158	65	853	140	56
v_{01}	1081	217	108	991	200	100	901	182	93
v_{02}	1168	299	184	1080	282	177	991	266	170
v_{03}	1309	433	312	1220	417	305	1132	401	298
v_{10}	1047	188	83	956	169	74	866	152	66
v_{11}	1110	242	129	1020	224	120	930	207	112
v_{12}	1195	322	204	1107	306	196	1019	290	190
v_{13}	1347	466	339	1259	450	333	1171	434	326
v_{20}	1071	209	102	981	191	93	891	174	85
v_{21}	1135	265	149	1046	247	141	957	230	133
v_{22}	1217	341	221	1129	325	214	1041	310	207
v_{23}	1370	487	358	1283	471	352	1195	456	345
v_{30}	1094	230	120	1004	212	112	915	195	104
v_{31}	1163	290	171	1074	273	163	985	256	156
v_{32}	1239	359	236	1151	343	229	1062	327	223
v_{33}	1420	531	398	1333	516	391	1245	501	385
D_{00}	P	P	P	P	P	P	P	P	P
D_{01}	P	P	P	P	P	NP	P	NP	NP
D_{02}	NP	NP	NP	NP	NP	NP	NP	NP	NP
D_{03}	NP	NP	NP	NP	NP	NP	NP	NP	NP
D_{10}	P	P	P	P	P	P	P	P	P
D_{11}	P	P	P	P	P	P	P	P	P
D_{12}	NP	NP	NP	NP	NP	NP	NP	NP	NP
D_{13}	NP	NP	NP	NP	NP	NP	NP	NP	NP
D_{20}	P	P	P	P	P	P	P	P	P
D_{21}	P	P	P	P	P	P	P	P	P
D_{22}	NP	NP	NP	NP	NP	NP	NP	NP	NP
D_{23}	NP	NP	NP	NP	NP	NP	NP	NP	NP
D_{30}	P	P	P	P	P	P	P	P	P
D_{31}	P	P	P	P	P	P	P	P	P
D_{32}	P	NP	NP	P	NP	NP	P	NP	NP
D_{33}	NP	NP	NP	NP	NP	NP	NP	NP	NP

Table 5.10. Optimal strategies when the second-order MDP is used.

| | $d = 3$ | | | $d = 4$ | | | $d = 5$ | | |
	$\alpha = 0.99$	$\alpha = 0.95$	$\alpha = 0.90$	$\alpha = 0.99$	$\alpha = 0.95$	$\alpha = 0.90$	$\alpha = 0.99$	$\alpha = 0.95$	$\alpha = 0.90$
x_0	14731	4277	2858	13572	4148	2825	13224	4093	2791
v_{00}	763	124	50	690	117	49	670	115	48
v_{01}	811	167	87	739	159	86	717	156	84
v_{02}	902	251	164	830	243	162	809	240	160
v_{03}	1044	386	293	972	378	290	951	375	288
v_{10}	776	135	59	703	127	57	682	124	55
v_{11}	841	191	107	768	182	105	745	179	103
v_{12}	930	275	184	858	267	182	836	263	180
v_{13}	1083	420	321	1012	412	319	990	409	317
v_{20}	801	158	79	728	150	77	707	146	74
v_{21}	867	214	127	794	206	124	771	201	121
v_{22}	953	295	202	881	287	200	859	284	198
v_{23}	1107	442	340	1035	434	338	1014	430	336
v_{30}	825	179	97	752	171	95	731	167	93
v_{31}	896	240	149	823	231	147	800	227	144
v_{32}	973	313	218	901	305	216	879	301	213
v_{33}	1158	487	381	1087	480	379	1065	476	377
D_{00}	P	P	NP	NP	NP	NP	NP	NP	NP
D_{01}	P	NP	NP	NP	NP	NP	NP	NP	NP
D_{02}	NP	NP	NP	NP	NP	NP	NP	NP	NP
D_{03}	NP	NP	NP	NP	NP	NP	NP	NP	NP
D_{10}	P	P	P	P	P	P	P	P	P
D_{11}	P	P	NP	P	NP	NP	P	NP	NP
D_{12}	NP	NP	NP	NP	NP	NP	NP	NP	NP
D_{13}	NP	NP	NP	NP	NP	NP	NP	NP	NP
D_{20}	P	P	P	P	P	P	P	P	P
D_{21}	P	P	P	P	P	P	P	P	P
D_{22}	NP	NP	NP	NP	NP	NP	NP	NP	NP
D_{23}	NP	NP	NP	NP	NP	NP	NP	NP	NP
D_{30}	P	P	P	P	P	P	P	P	P
D_{31}	P	P	P	P	P	P	P	P	P
D_{32}	P	NP	NP	P	NP	NP	NP	NP	NP
D_{33}	NP	NP	NP	NP	NP	NP	NP	NP	NP

6

Higher-order Markov Chains

6.1 Introduction

Data sequences or time series occur frequently in many real world applications. One of the most important steps in analyzing a data sequence (or time series) is the selection of an appropriate mathematical model for the data. Because it helps in predictions, hypothesis testing and rule discovery. A data sequence $\{X^{(n)}\}$ can be logically represented as a vector

$$(X^{(1)}, X^{(2)}, \cdots, X^{(T)}),$$

where T is the length of the sequence, and $X^{(i)} \in \mathrm{DOM}(A)$ ($1 \leq i \leq T$), associated with a defined semantic and a data type. In our context, we consider and assume other types used can be mapped to one of these two types. The domains of attributes associated with these two types are called numeric and categorical respectively. A numeric domain consists of real numbers. A domain $DOM(A)$ is defined as categorical if it is finite and unordered, e.g., for any $a, b \in DOM(A)$, either $a = b$ or $a \neq b$, see for instance [102]. Numerical data sequences have been studied in detail, see for instance [33]. Mathematical tools such as Fourier transform and spectral analysis are employed frequently in the analysis of numerical data sequences. Many different time sequences models have been proposed and developed in the literatures, see for instance [33].

For categorical data sequences, there are many situations that one would like to employ higher-order Markov chain models as a mathematical tool, see for instance [2, 140, 147, 149, 174]. A number of applications can be found in the literatures [114, 149, 175, 207]. For example, in sales demand prediction, products are classified into several states: very high sales volume, high sales volume, standard, low sales volume and very low sales volume (categorical type: ordinal data). A higher-order Markov chain model has been used in fitting observed data and apply to the wind turbine design. Alignment of sequences (categorical type: nominal data) is an important topic in DNA sequence analysis. It involves searching of patterns in a DNA sequence of

huge size. In these applications and many others, one would like to
(i) characterize categorical data sequences for the purpose of comparison and classification; or
(ii) to model categorical data sequences and hence to make predictions in the control and planning process.
It has been shown that higher-order Markov chain models can be a promising approach for these purposes [114, 174, 175, 207].

The remainder of this chapter is organized as follows. In Section 6.2, we present the higher-order Markov chain model. Estimation methods for the model parameters are also discussed. In Section 6.3, the higher-order Markov chain model is applied to a number of applications such as DNA sequences, sales demand predictions and web page predictions. Further extension of the model is then discussed in Section 6.4. In Section 6.5, we apply the model to the Newsboy's problem, a classical problem in management sciences. Finally a summary is given in Section 6.6.

6.2 Higher-order Markov Chains

In the following, we assume that each data point $X^{(n)}$ in a categorical data sequence takes values in the set

$$M \equiv \{1, 2, \ldots, m\}$$

and m is finite, i.e., a sequence has m possible categories or states. The conventional model for a k-th order Markov chain has $(m-1)m^k$ model parameters. The major problem in using such kind of model is that the number of parameters (the transition probabilities) increases exponentially with respect to the order of the model. This large number of parameters discourages people from using a higher-order Markov chain directly.

In [174], Raftery proposed a higher-order Markov chain model which involves only one additional parameter for each extra lag. The model can be written as follows:

$$P(X^{(n)} = j_0 \mid X^{(n-1)} = j_1, \ldots, X^{(n-k)} = j_k) = \sum_{i=1}^{k} \lambda_i q_{j_0 j_i} \qquad (6.1)$$

where

$$\sum_{i=1}^{k} \lambda_i = 1$$

and $Q = [q_{ij}]$ is a transition matrix with column sums equal to one, such that

$$0 \leq \sum_{i=1}^{k} \lambda_i q_{j_0 j_i} \leq 1, \quad j_0, j_i \in M. \qquad (6.2)$$

The constraint in (6.2) is to guarantee that the right-hand-side of (6.1) is a probability distribution. The total number of independent parameters in this model is of size $(k + m^2)$. Raftery proved that (6.1) is analogous to the standard AR(n) model in the sense that each additional lag, after the first is specified by a single parameter and the autocorrelations satisfy a system of linear equations similar to the Yule-Walker equations. Moreover, the parameters $q_{j_0 j_i}$ and λ_i can be estimated numerically by maximizing the log-likelihood of (6.1) subjected to the constraints (6.2). However, this approach involves solving a highly non-linear optimization problem. The proposed numerical method neither guarantees convergence nor a global maximum even if it converges.

6.2.1 The New Model

In this subsection, we extend Raftery's model [174] to a more general higher-order Markov chain model by allowing Q to vary with different lags. Here we assume that the weight λ_i is non-negative such that

$$\sum_{i=1}^{k} \lambda_i = 1. \tag{6.3}$$

It should be noted that (6.1) can be re-written as

$$\mathbf{X}^{(n+k+1)} = \sum_{i=1}^{k} \lambda_i Q \mathbf{X}^{(n+k+1-i)} \tag{6.4}$$

where $\mathbf{X}^{(n+k+1-i)}$ is the probability distribution of the states at time $(n + k + 1 - i)$. Using (6.3) and the fact that Q is a transition probability matrix, we note that each entry of $\mathbf{X}^{(n+k+1)}$ is in between 0 and 1, and the sum of all entries is also equal to 1. In Raftery's model, it does not assume λ to be non-negative and therefore the additional constraints (6.2) should be added to guarantee that $\mathbf{X}^{(n+k+1)}$ is the probability distribution of the states.

Raftery's model in (6.4) can be generalized as follows:

$$\mathbf{X}^{(n+k+1)} = \sum_{i=1}^{k} \lambda_i Q_i \mathbf{X}^{(n+k+1-i)}. \tag{6.5}$$

The total number of independent parameters in the new model is $(k + km^2)$. We note that if

$$Q_1 = Q_2 = \ldots = Q_k$$

then (6.5) is just the Raftery's model in (6.4).

In the model we assume that $\mathbf{X}^{(n+k+1)}$ depends on $\mathbf{X}^{(n+i)}$ ($i = 1, 2, \ldots, k$) via the matrix Q_i and weight λ_i. One may relate Q_i to the i-step transition matrix of the process and we will use this idea to estimate Q_i. Here we assume that each Q_i is an non-negative stochastic matrix with column sums equal to one. Before we present our estimation method for the model parameters we first discuss some properties of our proposed model in the following proposition.

Proposition 6.1. *If Q_k is irreducible and $\lambda_k > 0$ such that*

$$0 \leq \lambda_i \leq 1 \quad \text{and} \quad \sum_{i=1}^{k} \lambda_i = 1$$

then the model in (6.5) has a stationary distribution $\bar{\mathbf{X}}$ when $n \to \infty$ independent of the initial state vectors $\mathbf{X}^{(0)}, \mathbf{X}^{(1)}, \ldots, \mathbf{X}^{(k-1)}$. The stationary distribution $\bar{\mathbf{X}}$ is also the unique solution of the following linear system of equations:

$$(I - \sum_{i=1}^{n} \lambda_i Q_i)\bar{\mathbf{X}} = \mathbf{0} \quad \text{and} \quad \mathbf{1}^T \bar{\mathbf{X}} = 1.$$

Here I is the m-by-m identity matrix (m is the number of possible states taken by each data point) and $\mathbf{1}$ is an $m \times 1$ vector of ones.

Proof. We first note that if $\lambda_k = 0$, then this is not an kth order Markov chain. Therefore without loss of generality, one may assume that $\lambda_k > 0$. Secondly if Q_k is not irreducible, then we consider the case that $\lambda_k = 1$ and in this case, clearly there is no unique stationary distribution for the system. Therefore Q_k is irreducible is a necessary condition for the existence of a unique stationary distribution.

Now let

$$\mathbf{Y}^{(n+k+1)} = (\mathbf{X}^{(n+k+1)}, \mathbf{X}^{(n+k)}, \ldots, \mathbf{X}^{(n+2)})^T$$

be an $nm \times 1$ vector. Then one may write

$$\mathbf{Y}^{(n+1)} = R\mathbf{Y}^{(n)}$$

where

$$R = \begin{pmatrix} \lambda_1 Q_1 & \lambda_2 Q_2 & \cdots & \lambda_{n-1}Q_{n-1} & \lambda_n Q_n \\ I & 0 & \cdots & 0 & 0 \\ 0 & I & 0 & & \vdots \\ \vdots & \ddots & \ddots & \ddots & 0 \\ 0 & \cdots & 0 & I & 0 \end{pmatrix} \tag{6.6}$$

is an $km \times km$ square matrix. We then define

$$\tilde{R} = \begin{pmatrix} \lambda_1 Q_1 & I & 0 & 0 & \cdots\cdots & 0 \\ \vdots & & 0 & I & 0 & & \vdots \\ \vdots & & 0 & 0 & \ddots & \ddots & \vdots \\ \vdots & & \vdots & \ddots & \ddots & \ddots & 0 \\ \lambda_{n-1}Q_{n-1} & \vdots & & & \ddots & \ddots & I \\ \lambda_n Q_n & 0 & \cdots & & \cdots\cdots & & 0 \end{pmatrix}. \tag{6.7}$$

We note that R and $\tilde{R}$ have the same characteristic polynomial in τ:

$$det[(-1)^{k-1}((\lambda_1 Q_1 - \tau I)\tau^{k-1} + \sum_{i=2}^{k} \lambda_i Q_i \tau^{k-i})].$$

Thus R and $\tilde{R}$ have the same set of eigenvalues.

It is clear that $\tilde{R}$ is an irreducible stochastic matrix with column sums equal to one. Then from Perron-Frobenius Theorem [11, p. 134], all the eigenvalues of $\tilde{R}$ (or equivalently R) lie in the interval $(0, 1]$ and there is exactly one eigenvalue equal to one. This implies that

$$\lim_{n \to \infty} \overbrace{R \ldots R}^{n} = \lim_{n \to \infty} (R)^n = \mathbf{V}\mathbf{U}^T$$

is a positive rank one matrix as R is irreducible. Therefore we have

$$\begin{aligned}
\lim_{n \to \infty} \mathbf{Y}^{(n+k+1)} &= \lim_{n \to \infty} (R)^n \mathbf{Y}^{(k+1)} \\
&= \mathbf{V}(\mathbf{U}^t \mathbf{Y}^{(k+1)}) \\
&= \alpha \mathbf{V}.
\end{aligned}$$

Here α is a positive number because $\mathbf{Y}_{k+1} \neq \mathbf{0}$ and is non-negative. This implies that $X^{(n)}$ also tends to a stationary distribution as t goes to infinity. Hence we have

$$\lim_{n \to \infty} \mathbf{X}^{(n+k+1)} = \lim_{n \to \infty} \sum_{i=1}^{k} \lambda_i Q_i \mathbf{X}^{(n+k+1-i)}$$

and therefore we have

$$\bar{\mathbf{X}} = \sum_{i=1}^{k} \lambda_i Q_i \bar{\mathbf{X}}.$$

The stationary distribution vector $\bar{\mathbf{X}}$ satisfies

$$(I - \sum_{i=1}^{k} \lambda_i Q_i)\bar{\mathbf{X}} = \mathbf{0} \quad \text{with} \quad \mathbf{1}^T \bar{\mathbf{X}} = 1. \tag{6.8}$$

The normalization constraint is necessary as the matrix

$$(I - \sum_{i=1}^{k} \lambda_i Q_i)$$

has an one-dimensional null space. The result is then proved.

We remark that if some λ_i are equal to zero, one can rewrite the vector $\mathbf{Y}_{n+k+1}$ in terms of $\mathbf{X}_i$ where λ_i are nonzero. Then the model in (6.5) still has a stationary distribution $\bar{\mathbf{X}}$ when n goes to infinity independent of the initial state vectors. Moreover, the stationary distribution $\bar{\mathbf{X}}$ can be obtained by solving the corresponding linear system of equations with the normalization constraint.

6.2.2 Parameters Estimation

In this subsection, we present efficient methods to estimate the parameters Q_i and λ_i for $i = 1, 2, \ldots, k$. To estimate Q_i, one may regard Q_i as the i-step transition matrix of the categorical data sequence $\{X^{(n)}\}$. Given the categorical data sequence $\{X^{(n)}\}$, one can count the transition frequency $f_{jl}^{(i)}$ in the sequence from State l to State j in the i-step. Hence one can construct the i-step transition matrix for the sequence $\{X^{(n)}\}$ as follows:

$$F^{(i)} = \begin{pmatrix} f_{11}^{(i)} & \cdots\cdots & f_{m1}^{(i)} \\ f_{12}^{(i)} & \cdots\cdots & f_{m2}^{(i)} \\ \vdots & \vdots\quad\vdots & \vdots \\ f_{1m}^{(i)} & \cdots\cdots & f_{mm}^{(i)} \end{pmatrix}. \tag{6.9}$$

From $F^{(i)}$, we get the estimates for $Q_i = [q_{lj}^{(i)}]$ as follows:

$$\hat{Q}_i = \begin{pmatrix} \hat{q}_{11}^{(i)} & \cdots\cdots & \hat{q}_{m1}^{(i)} \\ \hat{q}_{12}^{(i)} & \cdots\cdots & \hat{q}_{m2}^{(i)} \\ \vdots & \vdots\quad\vdots & \vdots \\ \hat{q}_{1m}^{(i)} & \cdots\cdots & \hat{q}_{mm}^{(i)} \end{pmatrix} \tag{6.10}$$

where

$$\hat{q}_{lj}^{(i)} = \begin{cases} \dfrac{f_{lj}^{(i)}}{\displaystyle\sum_{l=1}^{m} f_{lj}^{(i)}} & \text{if } \displaystyle\sum_{l=1}^{m} f_{lj}^{(i)} \neq 0 \\[3ex] 0 & \text{otherwise.} \end{cases} \tag{6.11}$$

We note that the computational complexity of the construction of $F^{(i)}$ is of $O(L^2)$ operations, where L is the length of the given data sequence. Hence the total computational complexity of the construction of $\{F^{(i)}\}_{i=1}^{k}$ is of $O(kL^2)$ operations. Here k is the number of lags.

The following proposition shows that these estimators are unbiased.

Proposition 6.2. *The estimators in (6.11) satisfies*

$$E(f_{lj}^{(i)}) = q_{lj}^{(i)} E\left(\sum_{j=1}^{m} f_{lj}^{(i)} \right).$$

Proof. Let T be the length of the sequence, $[q_{lj}^{(i)}]$ be the i-step transition probability matrix and $\bar{X}_l$ be the steady state probability that the process is in state l. Then we have

$$E(f_{lj}^{(i)}) = T \cdot \bar{X}_l \cdot q_{lj}^{(i)}$$

and

$$E(\sum_{j=1}^{m} f_{lj}^{(i)}) = T \cdot \bar{X}_l \cdot (\sum_{j=1}^{m} q_{lj}^{(i)}) = T \cdot \bar{X}_l.$$

Therefore we have

$$E(f_{lj}^{(i)}) = q_{lj}^{(i)} \cdot E(\sum_{j=1}^{m} f_{lj}^{(i)}).$$

In some situations, if the sequence is too short then $\hat{Q}_i$ (especially $\hat{Q}_k$) contains a lot of zeros (therefore $\hat{Q}_n$ may not be irreducible). However, this did not occur in the tested examples. Here we propose the second method for the parameter estimation. Let $\mathbf{W}^{(i)}$ be the probability distribution of the i-step transition sequence, then another possible estimation for Q_i can be $\mathbf{W}^{(i)}\mathbf{1}^t$. We note that if $\mathbf{W}^{(i)}$ is a positive vector, then $\mathbf{W}^{(i)}\mathbf{1}^t$ will be a positive matrix and hence an irreducible matrix.

Proposition 6.1 gives a sufficient condition for the sequence $\mathbf{X}^{(n)}$ to converge to a stationary distribution $\mathbf{X}$. Suppose $\mathbf{X}^{(n)} \to \bar{\mathbf{X}}$ as n goes to infinity then $\bar{\mathbf{X}}$ can be estimated from the sequence $\{X^{(n)}\}$ by computing the proportion of the occurrence of each state in the sequence and let us denote it by $\hat{\mathbf{X}}$. From (6.8) one would expect that

$$\sum_{i=1}^{k} \lambda_i \hat{Q}_i \hat{\mathbf{X}} \approx \hat{\mathbf{X}}. \tag{6.12}$$

This suggests one possible way to estimate the parameters

$$\lambda = (\lambda_1, \ldots, \lambda_k)$$

as follows. One may consider the following minimization problem:

$$\min_{\lambda} || \sum_{i=1}^{k} \lambda_i \hat{Q}_i \hat{\mathbf{X}} - \hat{\mathbf{X}}||$$

subject to

$$\sum_{i=1}^{k} \lambda_i = 1, \quad \text{and} \quad \lambda_i \geq 0, \quad \forall i.$$

Here $||.||$ is certain vector norm. In particular, if $||.||_\infty$ is chosen, we have the following minimization problem:

$$\min_{\lambda} \max_{l} \left| \left[\sum_{i=1}^{k} \lambda_i \hat{Q}_i \hat{\mathbf{X}} - \hat{\mathbf{X}} \right]_l \right|$$

subject to

$$\sum_{i=1}^{k} \lambda_i = 1, \quad \text{and} \quad \lambda_i \geq 0, \quad \forall i.$$

Here $[\cdot]_l$ denotes the lth entry of the vector. The constraints in the optimization problem guarantee the existence of the stationary distribution $\mathbf{X}$. Next we see that the above minimization problem can be formulated as a linear programming problem:

$$\min_{\lambda} \; w$$

subject to

$$\begin{pmatrix} w \\ w \\ \vdots \\ w \end{pmatrix} \geq \hat{\mathbf{X}} - \left[\hat{Q}_1 \hat{\mathbf{X}} \mid \hat{Q}_2 \hat{\mathbf{X}} \mid \cdots \mid \hat{Q}_n \hat{\mathbf{X}} \right] \begin{pmatrix} \lambda_1 \\ \lambda_2 \\ \vdots \\ \lambda_n \end{pmatrix},$$

$$\begin{pmatrix} w \\ w \\ \vdots \\ w \end{pmatrix} \geq -\hat{\mathbf{X}} + \left[\hat{Q}_1 \hat{\mathbf{X}} \mid \hat{Q}_2 \hat{\mathbf{X}} \mid \cdots \mid \hat{Q}_n \hat{\mathbf{X}} \right] \begin{pmatrix} \lambda_1 \\ \lambda_2 \\ \vdots \\ \lambda_n \end{pmatrix},$$

$$w \geq 0, \quad \sum_{i=1}^{k} \lambda_i = 1, \quad \text{and} \quad \lambda_i \geq 0, \quad \forall i.$$

We can solve the above linear programming problem efficiently and obtain the parameters λ_i. In next subsection, we will demonstrate the estimation method by a simple example.

Instead of solving an min-max problem, one can also choose the $||.||_1$ and formulate the following minimization problem:

$$\min_{\lambda} \sum_{l=1}^{m} \left| \left[\sum_{i=1}^{k} \lambda_i \hat{Q}_i \hat{\mathbf{X}} - \hat{\mathbf{X}} \right]_l \right|$$

subject to

$$\sum_{i=1}^{k} \lambda_i = 1, \quad \text{and} \quad \lambda_i \geq 0, \quad \forall i.$$

The corresponding linear programming problem is given as follows:

$$\min_{\lambda} \sum_{l=1}^{m} w_l$$

subject to

$$
\begin{pmatrix} w_1 \\ w_2 \\ \vdots \\ w_m \end{pmatrix} \geq \hat{\mathbf{X}} - \left[\hat{Q}_1 \hat{\mathbf{X}} \mid \hat{Q}_2 \hat{\mathbf{X}} \mid \cdots \mid \hat{Q}_k \hat{\mathbf{X}} \right] \begin{pmatrix} \lambda_1 \\ \lambda_2 \\ \vdots \\ \lambda_k \end{pmatrix},
$$

$$
\begin{pmatrix} w_1 \\ w_2 \\ \vdots \\ w_m \end{pmatrix} \geq -\hat{\mathbf{X}} + \left[\hat{Q}_1 \hat{\mathbf{X}} \mid \hat{Q}_2 \hat{\mathbf{X}} \mid \cdots \mid \hat{Q}_k \hat{\mathbf{X}} \right] \begin{pmatrix} \lambda_1 \\ \lambda_2 \\ \vdots \\ \lambda_k \end{pmatrix},
$$

$$
w_i \geq 0, \quad \forall i, \quad \sum_{i=1}^{k} \lambda_i = 1, \quad \text{and} \quad \lambda_i \geq 0, \quad \forall i.
$$

In the above linear programming formulation, the number of variables is equal to k and the number of constraints is equal to $(2m + 1)$. The complexity of solving a linear programming problem is $O(k^3 L)$ where n is the number of variables and L is the number of binary bits needed to store all the data (the constraints and the objective function) of the problem [91].

We remark that other norms such as $||.||_2$ can also be considered. In this case, it will result in a quadratic programming problem. It is known that in approximating data by a linear function [79, p. 220], $||.||_1$ gives the most robust answer, $||.||_\infty$ avoids gross discrepancies with the data as much as possible and if the errors are known to be normally distributed then $||.||_2$ is the best choice. In the tested examples, we only consider the norms leading to solving linear programming problems.

6.2.3 An Example

We consider a sequence $\{X^{(n)}\}$ of three states ($m = 3$) given by

$$
\{1, 1, 2, 2, 1, 3, 2, 1, 2, 3, 1, 2, 3, 1, 2, 3, 1, 2, 1, 2\}. \tag{6.13}
$$

The sequence $\{X^{(n)}\}$ can be written in vector form

$$
X^{(1)} = (1, 0, 0)^T, \ X^{(2)} = (1, 0, 0)^T, \ X^{(3)} = (0, 1, 0)^T, \ \ldots, \ X^{(20)} = (0, 1, 0)^T.
$$

We consider $k = 2$, then from (6.13) we have the transition frequency matrices

$$
F^{(1)} = \begin{pmatrix} 1 & 3 & 3 \\ 6 & 1 & 1 \\ 1 & 3 & 0 \end{pmatrix} \quad \text{and} \quad F^{(2)} = \begin{pmatrix} 1 & 4 & 1 \\ 3 & 2 & 3 \\ 3 & 1 & 0 \end{pmatrix}. \tag{6.14}
$$

Therefore from (6.14) we have the i-step transition probability matrices ($i = 1, 2$) as follows:

$$\hat{Q}_1 = \begin{pmatrix} 1/8 & 3/7 & 3/4 \\ 3/4 & 1/7 & 1/4 \\ 1/8 & 3/7 & 0 \end{pmatrix} \quad \text{and} \quad \hat{Q}_2 = \begin{pmatrix} 1/7 & 4/7 & 1/4 \\ 3/7 & 2/7 & 3/4 \\ 3/7 & 1/7 & 0 \end{pmatrix} \quad (6.15)$$

and

$$\hat{X} = (\frac{2}{5}, \frac{2}{5}, \frac{1}{5})^T.$$

Hence we have

$$\hat{Q}_1\hat{X} = (\frac{13}{35}, \frac{57}{140}, \frac{31}{140})^T,$$

and

$$\hat{Q}_2\hat{X} = (\frac{47}{140}, \frac{61}{140}, \frac{8}{35})^T.$$

To estimate λ_i one can consider the optimization problem:

$$\min_{\lambda_1,\lambda_2} w$$

subject to

$$\begin{cases} w \geq \dfrac{2}{5} - \dfrac{13}{35}\lambda_1 - \dfrac{47}{140}\lambda_2 \\[2mm] w \geq -\dfrac{2}{5} + \dfrac{13}{35}\lambda_1 + \dfrac{47}{140}\lambda_2 \\[2mm] w \geq \dfrac{2}{5} - \dfrac{57}{140}\lambda_1 - \dfrac{61}{140}\lambda_2 \\[2mm] w \geq -\dfrac{2}{5} + \dfrac{57}{140}\lambda_1 + \dfrac{61}{140}\lambda_2 \\[2mm] w \geq \dfrac{1}{5} - \dfrac{31}{140}\lambda_1 - \dfrac{8}{35}\lambda_2 \\[2mm] w \geq -\dfrac{1}{5} + \dfrac{31}{140}\lambda_1 + \dfrac{8}{35}\lambda_2 \\[2mm] w \geq 0, \quad \lambda_1 + \lambda_2 = 1, \quad \lambda_1, \lambda_2 \geq 0. \end{cases}$$

The optimal solution is

$$(\lambda_1^*, \lambda_2^*, w^*) = (1, 0, 0.0286),$$

and we have the model

$$X^{(n+1)} = \hat{Q}_1 X^{(n)}. \quad (6.16)$$

We remark that if we do not specify the non-negativity of λ_1 and λ_2, the optimal solution becomes

$$(\lambda_1^{**}, \lambda_2^{**}, w^{**}) = (1.80, -0.80, 0.0157),$$

the corresponding model is

$$\mathbf{X}^{(n+1)} = 1.80\hat{Q}_1\mathbf{X}^{(n)} - 0.80\hat{Q}_2\mathbf{X}^{(n-1)}. \tag{6.17}$$

Although w^{**} is less than w^*, the model (6.17) is not suitable. It is easy to check that

$$1.80\hat{Q}_1 \begin{pmatrix} 1 \\ 0 \\ 0 \end{pmatrix} - 0.80\hat{Q}_2 \begin{pmatrix} 0 \\ 1 \\ 0 \end{pmatrix} = \begin{pmatrix} -0.2321 \\ 1.1214 \\ 0.1107 \end{pmatrix},$$

therefore λ_1^{**} and λ_2^{**} are not valid parameters.

We note that if we consider the minimization problem:

$$\min_{\lambda_1,\lambda_2} w_1 + w_2 + w_3$$

subject to

$$\begin{cases} w_1 \geq \dfrac{2}{5} - \dfrac{13}{35}\lambda_1 - \dfrac{47}{140}\lambda_2 \\[2mm] w_1 \geq -\dfrac{2}{5} + \dfrac{13}{35}\lambda_1 + \dfrac{47}{140}\lambda_2 \\[2mm] w_2 \geq \dfrac{2}{5} - \dfrac{57}{140}\lambda_1 - \dfrac{61}{140}\lambda_2 \\[2mm] w_2 \geq -\dfrac{2}{5} + \dfrac{57}{140}\lambda_1 + \dfrac{61}{140}\lambda_2 \\[2mm] w_3 \geq \dfrac{1}{5} - \dfrac{31}{140}\lambda_1 - \dfrac{9}{35}\lambda_2 \\[2mm] w_3 \geq -\dfrac{1}{5} + \dfrac{31}{140}\lambda_1 + \dfrac{9}{35}\lambda_2 \\[2mm] w_1, w_2, w_3 \geq 0, \quad \lambda_1 + \lambda_2 = 1, \quad \lambda_1, \lambda_2 \geq 0. \end{cases}$$

The optimal solution is the same as the previous min-max formulation and is equal to

$$(\lambda_1^*, \lambda_2^*, w_1^*, w_2^*, w_3^*) = (1, 0, 0.0286, 0.0071, 0.0214).$$

6.3 Some Applications

In this section we apply our model to some data sequences. The data sequences are the DNA sequence and the sales demand data sequence. Given the state vectors $\mathbf{X}^{(i)}$, $i = n - k, n - k + 1, \ldots, k - 1$, the state probability distribution at time n can be estimated as follows:

$$\hat{\mathbf{X}}^{(n)} = \sum_{i=1}^{k} \lambda_i \hat{Q}_i \mathbf{X}^{(n-i)}.$$

In many applications, one would like to make use of the higher-order Markov chain models for the purpose of prediction. According to this state probability distribution, the prediction of the next state $\hat{X}^{(n)}$ at time n can be taken as the state with the maximum probability, i.e.,

$$\hat{X}^{(n)} = j, \quad \text{if } [\hat{\mathbf{X}}^{(n)}]_i \leq [\hat{\mathbf{X}}^{(n)}]_j, \quad \forall 1 \leq i \leq m.$$

To evaluate the performance and effectiveness of the higher-order Markov chain model, a prediction accuracy r is defined as

$$r = \frac{1}{T} \sum_{t=k+1}^{T} \delta_t,$$

where T is the length of the data sequence and

$$\delta_t = \begin{cases} 1, & \text{if } \hat{X}^{(t)} = X^{(t)} \\ 0, & \text{otherwise.} \end{cases}$$

Using the example in the previous section, two possible prediction rules can be drawn as follows:

$$\begin{cases} \hat{X}^{(n+1)} = 2, & \text{if } X^{(n)} = 1, \\ \hat{X}^{(n+1)} = 1, & \text{if } X^{(n)} = 2, \\ \hat{X}^{(n+1)} = 1, & \text{if } X^{(n)} = 3 \end{cases}$$

or

$$\begin{cases} \hat{X}^{(n+1)} = 2, & \text{if } X^{(n)} = 1, \\ \hat{X}^{(n+1)} = 3, & \text{if } X^{(n)} = 2, \\ \hat{X}^{(n+1)} = 1, & \text{if } X^{(n)} = 3. \end{cases}$$

The prediction accuracy r for the sequence in (6.13) is equal to 12/19 for both prediction rules. While the prediction accuracies of other rules for the sequence in (6.13) are less than the value 12/19.

Next we present other numerical results on different data sequences are discussed. In the following tests, we solve min-max optimization problems to determine the parameters λ_i of higher-order Markov chain models. However, we remark that the results of using the $||.||_1$ optimization problem as discussed in the previous section are about the same as that of using the min-max formulation.

6.3.1 The DNA Sequence

In order to determine whether certain short DNA sequence (a categorical data sequence of four possible categories: A,C,G and T) occurred more often than would be expected by chance, Avery [8] examined the Markovian structure of introns from several other genes in mice. Here we apply our model to the introns from the mouse αA-crystallin gene see for instance [175]. We compare our second-order model with the Raftery's second-order model. The model

Table 6.1. Prediction accuracy in the DNA sequence.

	2-state model	3-state model	4-state model
New Model	0.57	0.49	0.33
Raftery's Model	0.57	0.47	0.31
Random Chosen	0.50	0.33	0.25

parameters of the Raftery's model are given in [175]. The results are reported in Table 6.1.

The comparison is made with different grouping of states as suggested in [175]. In grouping states 1 and 3, and states 2 and 4 we have a 2-state model. Our model gives

$$\hat{Q}_1 = \begin{pmatrix} 0.5568 & 0.4182 \\ 0.4432 & 0.5818 \end{pmatrix},$$

$$\hat{Q}_2 = \begin{pmatrix} 0.4550 & 0.5149 \\ 0.5450 & 0.4851 \end{pmatrix}$$

$$\hat{\mathbf{X}} = (0.4858, 0.5142)^T, \quad \lambda_1 = 0.7529 \quad \text{and} \quad \lambda_2 = 0.2471.$$

In grouping states 1 and 3 we have a 3-state model. Our model gives

$$\hat{Q}_1 = \begin{pmatrix} 0.5568 & 0.3573 & 0.4949 \\ 0.2571 & 0.3440 & 0.2795 \\ 0.1861 & 0.2987 & 0.2256 \end{pmatrix},$$

$$\hat{Q}_2 = \begin{pmatrix} 0.4550 & 0.5467 & 0.4747 \\ 0.3286 & 0.2293 & 0.2727 \\ 0.2164 & 0.2240 & 0.2525 \end{pmatrix}$$

$$\hat{\mathbf{X}} = (0.4858, 0.2869, 0.2272)^T, \quad \lambda_1 = 1.0 \quad \text{and} \quad \lambda_2 = 0.0$$

If there is no grouping, we have a 4-state model. Our model gives

$$\hat{Q}_1 = \begin{pmatrix} 0.2268 & 0.2987 & 0.2274 & 0.1919 \\ 0.2492 & 0.3440 & 0.2648 & 0.2795 \\ 0.3450 & 0.0587 & 0.3146 & 0.3030 \\ 0.1789 & 0.2987 & 0.1931 & 0.2256 \end{pmatrix},$$

$$\hat{Q}_2 = \begin{pmatrix} 0.1891 & 0.2907 & 0.2368 & 0.2323 \\ 0.3814 & 0.2293 & 0.2773 & 0.2727 \\ 0.2532 & 0.2560 & 0.2305 & 0.2424 \\ 0.1763 & 0.2240 & 0.2555 & 0.2525 \end{pmatrix}$$

$$\hat{\mathbf{X}} = (0.2395, 0.2869, 0.2464, 0.2272)^T, \quad \lambda_1 = 0.253 \quad \text{and} \quad \lambda_2 = 0.747.$$

When using the expected errors (assuming that the next state is randomly chosen with equal probability for all states) as a reference, the percentage gain in effectiveness of using higher-order Markov chain models is in the 3-state model. In this case, our model also gives a better estimation when compared with Raftery's model. Raftery [174] considered using BIC to weight efficiency gained in terms of extra parameters used. This is important in his approach since his method requires to solve a highly non-linear optimization problem. The complexity of solving the optimization problem increases when there are many parameters to be estimated. We remark that our estimation method is quite efficient.

6.3.2 The Sales Demand Data

A large soft-drink company in Hong Kong presently faces an in-house problem of production planning and inventory control. A pressing issue that stands out is the storage space of its central warehouse, which often finds itself in the state of overflow or near capacity. The company is thus in urgent needs to study the interplay between the storage space requirement and the overall growing sales demand. There are product states due to the level of sales volume. The states include

    state 1: very slow-moving (very low sales volume);
    state 2: slow-moving;
    state 3: standard;
    state 4: fast-moving;
    state 5: very fast-moving (very high sales volume).

Such labellings are useful from both marketing and production planning points of view. For instance, in the production planning, the company can develop a dynamic programming (DP) model to recommend better production planning so as to minimize its inventory build-up, and to maximize the demand satisfaction as well. Since the number of alternatives at each stage (each day in the planning horizon) are very large (the number of products raised to the power of the number of production lines), the computational complexity of the DP model is enormous. A priority scheme based on the state (the level of sales volume) of the product is introduced to tackle this combinatorial problem, and therefore an effective and efficient production plan can be obtained. It is obvious that the accurate prediction of state (the level of sales volume) of the product is important in the production planning model.

In Figure 6.1 (Taken from [62]), we show that the states of four of the products of the soft-drink company for some sales periods. Here we employ higher-order Markov chain models to predict categories of these four products separately. For the new model, we consider a second-order ($n = 2$) model and use the data to estimate $\hat{Q}_i$ and λ_i ($i = 1, 2$). The results are reported in

Table 6.2. For comparison, we also study the first-order and the second-order full Markov chain model. Results shows the effectiveness of our new model. We also see from Figure 6.1 that the change of the states of the products A, B and D is more regular than that of the product C. We find in Table 6.2 that the prediction results for the products A, B and D are better than that of C.

Table 6.2. Prediction accuracy in the sales demand data.

	Product A	Product B	Product C	Product D
First-order Markov Chain Model	0.76	0.70	0.39	0.74
Second-order Markov Chain Model	0.79	0.78	0.51	0.83
New Model $(n = 2)$	0.78	0.76	0.43	0.78
Random Chosen	0.20	0.20	0.20	0.20

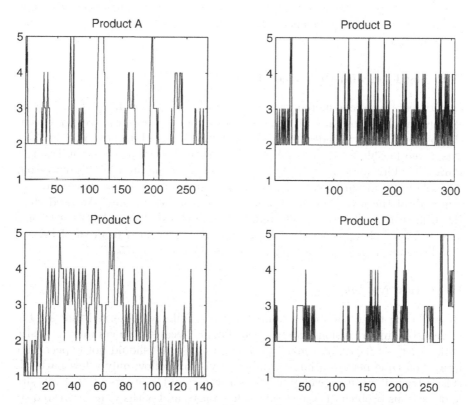

Fig. 6.1. The states of four products A,B,C and D.

6.3.3 Webpages Prediction

The Internet provides a rich environment for users to retrieve useful information. However, it is easy for a user to get lost in the ocean of information. One way to assist the user with their informational need is to predict a user's future request and use the prediction for recommendation. Recommendation systems reply on a prediction model to make inferences on users' interests based upon which to make recommendations. Examples are the WebWatcher [121] system and Letzia [141] system. Accurate prediction can potentially shorten the users' access times and reduce network traffic when the recommendation is handled correctly. In this subsection, we use a higher-order Markov chain model to exploit the information from web server logs for predicting users' actions on the web.

The higher-order Markov chain model is built on a web server log file. We consider the web server log file to be preprocessed into a collection of user sessions. Each session is indexed by a unique user ID and starting time [183]. Each session is a sequence of requests where each request corresponds to a visit to a web page. We represent each request as a state. Then each session is just a categorical data sequence. Moreover, we denote each Web page (state) by an integer.

Web Log Files and Preprocessing

Experiments were conducted on a real Web log file taken from the Internet. We first implemented a data preprocessing program to extract sessions from the log file. We downloaded two web log files from the Internet. The data set was a web log file from the EPA WWW server located at Research Triangle Park, NC. This log contained 47748 transactions generated in 24 hours from 23:53:25 EDT, August 29, to 23:53:07, August 30, 1995. In preprocessing, we removed all the invalid requests and the requests for images. We used Host ID to identify visitors and a 30 minutes time threshold to identify sessions. 428 sessions of lengths between 16 and 20 were identified from the EPA log file. The total number of web pages (states) involved is 3753.

Prediction Models

By exploring the session data from the web log file, we observed that a large number of similar sessions rarely exist. This is because in a complex web site with variety of pages, and many paths and links, one should not expect that in a given time period, a large number of visitors follow only a few paths. If this is true, it would mean that the structure and contents of the web site had a serious problem. Because only a few pages and paths were interested by the visitors. In fact, most web site designers expect that the majority of their pages, if not every one, are visited and paths followed (equally) frequently. The first and the second step transition matrices of all sessions are very sparse in our case. In fact, there are 3900 and 4747 entries in the first and the second step transition matrices respectively. Nonzero entries only contain about 0.033% in the total elements of the first and the second step transition matrices.

Based on these observations, if we directly use these transition matrices to build prediction models, they may not be effective. Since the number of pages (states) are very large, the prediction probability for each page may be very low. Moreover, the computational work for solving the linear programming problem in the estimation of λ_i are also high since the number of constraints in the linear programming problem depends on the number of pages (states). Here we propose to use clustering algorithms [114] to cluster the sessions. The idea is to form a transition probability matrix for each session, to construct the distance between two sessions based on the Frobenius norm (See Definition 1.40 of Chapter one) of the difference of their transition probability matrices, and then to use k-means algorithm to cluster the sessions. As a result of the cluster analysis, the web page cluster can be used to construct a higher-order Markov chain model. Then we prefetch those web documents that are close to a user-requested document in a Markov chain model.

We find that there is a clear similarity among these sessions in each cluster for the EPA log file. As an example, we show in Figure 6.2 (Taken from [62]) that the first, the second and the third step transition probability matrices of a cluster in EPA log file. There are 70 pages involved in this cluster. Non-zero entries contain about 1.92%, 2.06% and 2.20% respectively in the total elements of the first, the second and the third step transition matrices. Usually, the prediction of the next web page is based on the current page and the previous few pages [1]. Therefore, we use a third-order model ($n = 3$) and consider the first, the second and the third transition matrices in the construction of the Markov chain model. After we find the transition matrices, we determine λ_i and build our new higher-order Markov chain model for each cluster. For the above mentioned cluster, its corresponding λ_1, λ_2 and λ_3 are 0.4984, 0.4531 and 0.0485 respectively. The parameters show that the prediction of the next web page strongly depends on the current and the previous pages.

Prediction Results

We then present the prediction results for the EPA log file. We perform clustering based on their transition matrices and parameters. Sixteen clusters are found experimentally based on their average within-cluster distance. Therefore sixteen third-order Markov chain model for these clusters are determined for the prediction of user-request documents. For comparison, we also compute the first-order Markov chain model for each cluster. Totally, there are 6255 web documents for the prediction test. We find the prediction accuracy of our method is about 77%, but the prediction accuracy of using the first-order full Markov chain model is only 75%. Results show an improvement in the prediction. We have applied these prediction results to the problem of

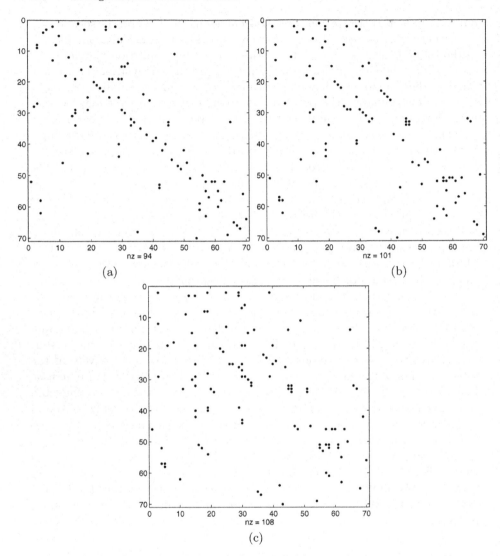

Fig. 6.2. The first (a), second (b), third (c) step transition matrices.

integrated web caching and prefetching [212]. The slight increase of the prediction accuracy can enhance a prefetching engine. Experimental results in
[212] show that the resultant system outperforms web systems that are based
on caching alone.

6.4 Extension of the Model

In this section, we consider an extension of the higher-order Markov chain model, Ching et al. [71]. The higher-order Markov chain model (6.5):

$$\mathbf{X}_{n+k+1} = \sum_{i=1}^{k} \lambda_i Q_i \mathbf{X}_{n+k+1-i}$$

can be further generalized by replacing the constraints

$$0 \leq \lambda_i \leq 1, \quad i = 1, 2, \ldots, k \quad \text{and} \quad \sum_{i=1}^{k} \lambda_i = 1$$

by

$$0 \leq \sum_{i=1}^{k} \lambda_i q_{j_0 j_i}^{(i)} \leq 1, \quad j_0, j_i \in \mathcal{M} \quad \text{and} \quad \sum_{i=1}^{k} \lambda_i = 1.$$

We expect this new model will have better prediction accuracy when appropriate order of model is used.

Next we give the sufficient condition for the proposed model to be stationary. Similar to the proof in [174], it can be shown that

Proposition 6.3. *Suppose that* $\{X^{(n)}, n \in N\}$ *is defined by (6.5) where the constraints* $0 \leq \lambda \leq 1$ *are replaced by*

$$0 < \sum_{i=1}^{k} \lambda_i q_{j_0 j_i}^{(i)} \leq 1,$$

then the model (6.5) has a stationary distribution $\bar{\mathbf{X}}$ *when* $n \to \infty$ *independent of the initial state vectors*

$$(\mathbf{X}^{(0)}, \mathbf{X}^{(1)}, \ldots, \mathbf{X}^{(k-1)}).$$

The stationary distribution $\bar{\mathbf{X}}$ *is also the unique solution of the linear system of equations:*

$$(I - \sum_{i=1}^{k} \lambda_i Q_i)\bar{\mathbf{X}} = 0 \quad \text{and} \quad \mathbf{1}^T \bar{\mathbf{X}} = 1.$$

We can use the method in Section 6.2.2 to estimate the parameters Q_i. For λ_i, the linear programming formulation can be considered as follows. In view of Proposition 6.3, suppose the model is stationary then we have a stationary distribution $\bar{\mathbf{X}}$. Then $\bar{\mathbf{X}}$ can be estimated from the observed sequence $\{X^{(s)}\}$ by computing the proportion of the occurrence of each state in the sequence. In Section 6.2.2, it suggests one possible way to estimate the parameters

$$\lambda = (\lambda_1, \ldots, \lambda_k)$$

as follows. In view of (6.12) one can consider the following optimization problem:

$$\min_{\lambda} \left\| \sum_{i=1}^{k} \lambda_i \hat{Q}_i \hat{\mathbf{X}} - \hat{\mathbf{X}} \right\|_{\infty} = \min_{\lambda} \max_{j} \left[\left\| \sum_{i=1}^{k} \lambda_i \hat{Q}_i \hat{\mathbf{X}} - \hat{\mathbf{X}} \right\| \right]_j$$

subject to

$$\sum_{i=1}^{k} \lambda_i = 1,$$

and

$$0 \le \sum_{i=1}^{k} \lambda_i q_{j_0 j_i}^{(i)} \le 1, \quad j_0, j_i \in \mathcal{M}.$$

Here $[\cdot]_j$ denotes the jth entry of the vector. We see that the above optimization problem can be re-formulated as a linear programming problem as stated in the previous section. Instead of solving a min-max problem, one can also formulate the l_1-norm optimization problem In these linear programming problems, we note that the number of variables is equal to k and the number of constraints is equal to $(2m^{k+1}+2m+1)$. With the following proposition (see also [175]), we can reduce number of constraints to $(4m + 1)$ if we formulate the estimation problem as a nonlinear programming.

Proposition 6.4. *The constraints*

$$0 \le \sum_{i=1}^{k} \lambda_i q_{j_0 j_i}^{(i)} \le 1, \quad j_0, j_i \in \mathcal{M}$$

are equivalent to

$$\sum_{i=1}^{k} \left(\max\{\lambda_i, 0\} \min_{j_i}\{q_{j_0 j_i}^{(i)}\} - \max\{-\lambda_i, 0\} \max_{j_i}\{q_{j_0 j_i}^{(i)}\} \right) \ge 0 \qquad (6.18)$$

and

$$\sum_{i=1}^{k} \left(\max\{\lambda_i, 0\} \max_{j_i}\{q_{j_0 j_i}^{(i)}\} - \max\{-\lambda_i, 0\} \min_{j_i}\{q_{j_0 j_i}^{(i)}\} \right) \le 1 \qquad (6.19)$$

Proof. We prove the first part of the inequality. If inequality (6.18) holds, then

$$\sum_{i=1}^{k} \lambda_i q_{j_0 j_i}^{(i)} = \sum_{\lambda_i \ge 0} \lambda_i q_{j_0 j_i}^{(i)} + \sum_{\lambda_i < 0} \lambda_i q_{j_0 j_i}^{(i)}$$

$$\ge \sum_{\lambda_i \ge 0} \lambda_i \min_{j_i}\{q_{j_0 j_i}^{(i)}\} + \sum_{\lambda_i < 0} \lambda_i \max_{j_i}\{q_{j_0 j_i}^{(i)}\}$$

$$\ge 0.$$

Conversely, we assume that

$$\forall j_0, j_i \in \mathcal{M}, \quad \sum_{i=1}^{k} \lambda_i q_{j_0 j_i}^{(i)} \geq 0.$$

Suppose

$$\min_{j_i}\{q_{j_0 j_i}^{(i)}\} = q_{j_0 j_{i_0}}^{(i)}$$

and

$$\max_{j_i}\{q_{j_0 j_i}^{(i)}\} = q_{j_0 j_{i_1}}^{(i)}$$

then

$$\sum_{\lambda_i \geq 0} \lambda_i \min_{j_i}\{q_{j_0 j_i}^{(i)}\} + \sum_{\lambda_i < 0} \lambda_i \max_{j_i}\{q_{j_0 j_i}^{(i)}\} = \sum_{\lambda_i \geq 0} \lambda_i q_{j_0 j_{i_0}}^{(i)} + \sum_{\lambda_i < 0} \lambda_i q_{j_0 j_{i_1}}^{(i)} \geq 0.$$

This is equivalent to (6.18). One can use similar method to prove the second part and hence the proof.

In the following, we give a simple example to demonstrate our estimation methods. We consider a sequence $\{X^{(t)}\}$ of two states ($m = 2$) given by

$$\{1, 1, 2, 2, 1, 2, 2, 1, 2, 2, 1, 2, 2, 1, 2, 2, 1, 2, 2, 2\}. \tag{6.20}$$

The sequence $\{X^{(t)}\}$ can be written in vector form

$$X^{(1)} = (1, 0)^T, \quad X^{(2)} = (1, 0)^T, \quad X^{(3)} = (0, 1)^T, \quad \ldots \quad, X^{(20)} = (0, 1)^T.$$

We consider $k = 2, 3, 4$, then from (6.20) we have the transition frequency matrices

$$F^{(1)} = \begin{pmatrix} 1 & 5 \\ 6 & 7 \end{pmatrix}, \quad F^{(2)} = \begin{pmatrix} 0 & 5 \\ 7 & 6 \end{pmatrix}, \tag{6.21}$$

$$F^{(3)} = \begin{pmatrix} 5 & 0 \\ 2 & 10 \end{pmatrix}, \quad F^{(4)} = \begin{pmatrix} 1 & 4 \\ 5 & 6 \end{pmatrix}. \tag{6.22}$$

Therefore from (6.21) we have the i-step transition matrices ($i = 1, 2, 3, 4$) as follows:

$$\hat{Q}_1 = \begin{pmatrix} 1/7 & 5/12 \\ 6/7 & 7/12 \end{pmatrix}, \quad \hat{Q}_2 = \begin{pmatrix} 0 & 5/11 \\ 1 & 6/11 \end{pmatrix}, \tag{6.23}$$

$$\hat{Q}_3 = \begin{pmatrix} 5/7 & 0 \\ 2/7 & 1 \end{pmatrix}, \quad \hat{Q}_4 = \begin{pmatrix} 1/6 & 4/10 \\ 5/6 & 6/10 \end{pmatrix}. \tag{6.24}$$

and $\hat{X} = (0.35, 0.65)^T$. In this example, the model parameters can be obtained by solving a linear programming problem. It turns out that the parameters obtained are identical the same for both $\| \cdot \|_1$ and $\| \cdot \|_\infty$. We report the parameters for the case of $k = 2, 3, 4$. For $k = 2$, we have

$$(\lambda_1^*, \lambda_2^*) = (1.4583, -0.4583).$$

For $k = 3$, we have
$$(\lambda_1^*, \lambda_2^*, \lambda_3^*) = (1.25, 0, -0.25).$$

For $k = 4$, we have

$$(\lambda_1^*, \lambda_2^*, \lambda_3^*, \lambda_4^*) = (0, 0, -0.3043, 1.3043).$$

Next we present the numerical comparisons with the data set in the previous section, (let us denote it by "Sample") and also the DNA data set of 3-state sequence from the mouse αA-crystallin gene, (let us denote it by "DNA"). The length of the sequence of "Sample" is 20 and the length of the sequence of "DNA" is 1307. The results are reported in Tables 6.3 and 6.4 below.

We then present the χ^2 statistics method. From the observed data sequence, one can obtain the distribution of states

$$(O_1, O_2, \ldots, O_m).$$

From the model parameters Q_i and λ_i, by solving:

$$\mathbf{X} = \sum_{i=1}^{n} \lambda_i \hat{Q}_i \mathbf{X} \quad \text{with} \quad \mathbf{1}^T \mathbf{X} = 1$$

one can obtain the theoretical probability distribution of the states

$$(E_1, E_2, \ldots, E_m).$$

Then the χ^2 statistics is defined as

$$\chi^2 = L \sum_{i=1}^{m} \frac{(E_i - O_i)^2}{E_i}.$$

The smaller this value is the better the model will be.

We note that for the "Sample" data set, significant improvement in prediction accuracy is observed when the order is increased from 2 to 4. In this case, except the last state all the other states can be predicted correctly. For all the "DNA" data set, the best model is our new extended model with order 4, 3, 2 corresponding to 2-state, 3-state, 4-state sequence. For the 2-state and 3-state sequence, we can get much better prediction accuracy than the higher-order Markov chain in the previous section. For the 4-state sequence, we also can get the same prediction accuracy as the model in previous section.

Table 6.3. Prediction accuracy and χ^2 value.

$n = 2$	Sample (2-state)	DNA (2-state)
Extended Model ($\|.\|_\infty$)	0.3889 ($\chi^2 = 1.2672$)	0.5295 ($\chi^2 = 0.0000$)
Extended Model ($\|.\|_1$)	0.3889 ($\chi^2 = 1.2672$)	0.5295 ($\chi^2 = 0.0000$)
Ching's Model ($\|.\|_\infty$)	0.6842 ($\chi^2 = 3.1368$)	0.5295 ($\chi^2 = 0.0000$)
Ching's Model ($\|.\|_1$)	0.6842 ($\chi^2 = 3.1368$)	0.5295 ($\chi^2 = 0.0000$)
Randomly Chosen	0.5000	0.5000

$n = 3$	Sample (2-state)	DNA (2-state)
Extended Model ($\|.\|_\infty$)	0.3529 ($\chi^2 = 0.3265$)	0.5299 ($\chi^2 = 0.0000$)
Extended Model ($\|.\|_1$)	0.3529 ($\chi^2 = 0.3265$)	0.5299 ($\chi^2 = 0.0000$)
New Model ($\|.\|_\infty$)	0.6842 ($\chi^2 = 3.1368$)	0.5295 ($\chi^2 = 0.0000$)
New Model ($\|.\|_1$)	0.6842 ($\chi^2 = 3.1368$)	0.5295 ($\chi^2 = 0.0000$)
Randomly Chosen	0.5000	0.5000

$n = 4$	Sample (2-state)	DNA (2-state)
Extended Model ($\|.\|_\infty$)	0.9375 ($\chi^2 = 0.2924$)	0.5375 ($\chi^2 = 0.0000$)
New Model ($\|.\|_1$)	0.9375 ($\chi^2 = 0.2924$)	0.5372 ($\chi^2 = 0.0000$)
New Model ($\|.\|_\infty$)	0.6842 ($\chi^2 = 3.1368$)	0.5295 ($\chi^2 = 0.0000$)
New Model ($\|.\|_1$)	0.6842 ($\chi^2 = 3.1368$)	0.5295 ($\chi^2 = 0.0000$)
Randomly Chosen	0.5000	0.5000

Table 6.4. Prediction accuracy and χ^2 value.

$n = 2$	DNA (3-state)	DNA (4-state)
Extended Model ($\|.\|_\infty$)	0.4858 ($\chi^2 = 7.09E-4$)	0.3303 ($\chi^2 = 0.0030$)
Extended Model ($\|.\|_1$)	0.4858 ($\chi^2 = 7.09E-4$)	0.3287 ($\chi^2 = 0.0022$)
New Model ($\|.\|_\infty$)	0.4858 ($\chi^2 = 7.09E-4$)	0.3303 ($\chi^2 = 0.0030$)
New Model ($\|.\|_1$)	0.4858 ($\chi^2 = 7.09E-4$)	0.3287 ($\chi^2 = 0.0022$)
Randomly Chosen	0.3333	0.2500

$n = 3$	DNA (3-state)	DNA (4-state)
Extended Model ($\|.\|_\infty$)	0.4946 ($\chi^2 = 4.24E-4$)	0.3083 ($\chi^2 = 0.0039$)
Extended Model ($\|.\|_1$)	0.4893 ($\chi^2 = 8.44E-5$)	0.3282 ($\chi^2 = 0.0050$)
New Model ($\|.\|_\infty$)	0.4858 ($\chi^2 = 7.09E-4$)	0.3277 ($\chi^2 = 0.0032$)
New Model ($\|.\|_1$)	0.4858 ($\chi^2 = 7.09E-4$)	0.3282 ($\chi^2 = 0.0052$)
Randomly Chosen	0.3333	0.2500

$n = 4$	Sample (3-state)	DNA (4-state)
Extended Model ($\|.\|_\infty$)	0.4666 ($\chi^2 = 1.30E-4$)	0.3085 ($\chi^2 = 0.0039$)
Extended Model ($\|.\|_1$)	0.4812 ($\chi^2 = 4.55E-5$)	0.3031 ($\chi^2 = 0.0047$)
New Model ($\|.\|_\infty$)	0.4858 ($\chi^2 = 7.09E-4$)	0.3277 ($\chi^2 = 0.0032$)
New Model ($\|.\|_1$)	0.4858 ($\chi^2 = 7.09E-4$)	0.3285 ($\chi^2 = 0.0044$)
Randomly Chosen	0.3333	0.2500

6.5 Newboy's Problems

The Newsboy's problem is a well-known classical problem in management science [158] and it can be described as follows. A newsboy start selling newspaper every morning. The cost of each newspaper remaining unsold at the end of the day is C_o (overage cost) and the cost of each unsatisfied demand is C_s (shortage cost). Suppose that the probability distribution function of the demand D is given by

$$\text{Prob } (D = d) = p_d \geq 0, \quad d = 1, 2, \ldots, m. \tag{6.25}$$

The objective here is to determine the best amount r^* of newspaper to be ordered such that the expected cost is minimized. To write down the expected long-run cost for a given amount of order size r we have the following two cases.

(i) If the demand $d < r$, then the cost will be $(r - d)C_o$ and
(ii) if the demand $d > r$, then the cost will be $(d - r)C_s$.

Therefore the expected cost when the order size is r is given by

$$\mathcal{E}(r) = \underbrace{C_o \sum_{d=1}^{r} (r - d)p_i}_{\text{Expected Overage Cost}} + \underbrace{C_s \sum_{d=r+1}^{m} (d - r)p_i}_{\text{Expected Shortage Cost}}. \tag{6.26}$$

Let us define the cumulative probability function of the demand D as follows:

$$F(d) = \sum_{i=1}^{d} p_i = \text{Prob } (D \leq d) \quad \text{for} \quad d = 1, 2, \ldots, m. \tag{6.27}$$

We have the following results.

Proposition 6.5.

$$\mathcal{E}(r) - \mathcal{E}(r + 1) = C_s - (C_o + C_s)F(r) \tag{6.28}$$

and

$$\mathcal{E}(r) - \mathcal{E}(r - 1) = -C_s + (C_o + C_s)F(r - 1). \tag{6.29}$$

By using the above lemma and making use of the fact that $F(r)$ is monotonically increasing in r, we have the following proposition.

Proposition 6.6. *The optimal order size r^* is the one which satisfies*

$$F(r^* - 1) < \frac{C_s}{C_s + C_o} \leq F(r^*). \tag{6.30}$$

6.5.1 A Markov Chain Model for the Newsboy's Problem

One can further generalize the Newsboy's problem as follows. Suppose that the demand is governed by a Markov chain, i.e., the demand tomorrow depends on the demand today. Again the demand has m possible states. We shall order the states in increasing order. The demand at time t is said to be in state i if the demand is i and is denoted by the vector

$$X_t = (0,\ldots,0,\ \underbrace{1}_{i\text{th entry}},0\ldots,0)^T.$$

We let Q (an $m \times m$ matrix) to be the transition probability matrix of the Markov process of the demand. Therefore we have

$$X_{t+1} = QX_t.$$

Here we assume that Q is irreducible and hence the stationary probability distribution S exists, i.e.

$$\lim_{t\to\infty} X_t = S = (s_1, s_2, \ldots, s_m)^T.$$

Now we let $r_j \in \{1, 2, \ldots, m\}$ be the size of the next order given that the current demand is j and $C(r_j, i)$ be the cost of the situation that the size of order is r_j and the actual next demand is i. We note that $C(r_j, i)$ is a more general cost than the one in (6.26). Clearly the optimal ordering policy depends on the state of the current demand because the demand probability distribution in the next period depends on the state of the current demand. The expected cost is then given by

$$\mathcal{E}(\{r_1, r_2, \ldots, r_m\}) = \sum_{j=1}^{m} s_j \times \left(\sum_{i=1}^{m} C(r_j, i) q_{ij} \right) \tag{6.31}$$

where $q_{ij} = [Q]_{ij}$ is the transition probability of the demand from the state j to the state i. In other words, q_{ij} is the probability that the next demand will be in state i given that the current demand is in state j. The optimal ordering policy

$$(r_1^*, r_2^*, \ldots, r_m^*)$$

is the one which minimizes (6.31). We observe that if the current demand is j, then we only need to choose the ordering size r_j to minimize the expected cost. Since

$$\min_{r_j} \mathcal{E}(\{r_1, r_2, \ldots, r_m\}) = \sum_{j=1}^{m} s_j \times \left(\min_{r_j} \sum_{i=1}^{m} C(r_j, i) q_{ij} \right), \tag{6.32}$$

the optimal ordering size r_j^* can be obtained by solving

$$\min_{r_j} \sum_{i=1}^{m} C(r_j, i) q_{ij}. \tag{6.33}$$

By using Proposition 6.6, we have

Proposition 6.7. *If*

$$C(r_j, i) = \begin{cases} C_o(r_j - i) & \text{if } r_j \geq i \\ C_s(i - r_j) & \text{if } r_j < i \end{cases} \tag{6.34}$$

and let

$$F_j(k) = \sum_{i=1}^{k} q_{ij}$$

then the optimal ordering size r_j^* satisfies

$$F_j(r_j^* - 1) < \frac{C_s}{C_s + C_o} \leq F_j(r_j^*).$$

We remark that one has to estimate q_{ij} before one can apply the Markov chain model. We will propose an estimation method for q_{ij} as discussed in the previous section. We note that when $q_{ij} = q_i$ for $i, j = 1, 2, \ldots, m$, (the demand distribution is stationary and independent of the current demand state) then the Markov Newsboy model described above reduces to the classical Newsboy's problem. Let us consider an example to demonstrate that the extension to a Markov chain model is useful and important.

Example 6.8. Suppose that the demand $(1, 2, \ldots, 2k)$ $(m = 2k)$ follows a Markov process with the transition probability matrix Q of size $2k \times 2k$ given by

$$Q = \begin{pmatrix} 0 & 0 & \cdots & 0 & 1 \\ 1 & 0 & \ddots & & 0 \\ 0 & 1 & 0 & \ddots & \vdots \\ \vdots & \ddots & \ddots & \ddots & 0 \\ 0 & \cdots & 0 & 1 & 0 \end{pmatrix} \tag{6.35}$$

and the cost is given in (6.34) with $C_o = C_s$. Clearly the next demand can be determined certainly by the state of the current demand, and hence the optimal expected cost is equal to zero when the Markov chain model is used. When the classical Newsboy model is used, we note that the stationary distribution of Q is given by

$$\frac{1}{2k}(1, 1, \ldots, 1)^T.$$

The optimal ordering size is equal to k by Proposition 6.6 and therefore the optimal expected cost is $C_o k$.

According to this example, it is obvious that the more "information" one can extract from the demand sequence, the better the model will be and hence the better the optimal ordering policy one can obtain. Therefore it is natural for one to consider a higher-order Markov chain model. The only obstacle here is the huge number of states and parameters. We employ a higher-order Markov chain model that can cope with the difficulty.

Let us study the optimal ordering policy for this higher-order Markov chain model. Define the set

$$\Phi = \{G = (j_1, j_2, \ldots, j_n)^T \mid j_k \in \{1, 2, \ldots, m\} \text{ for } k = 1, 2, \ldots, n\}.$$

let

$$p_{i,G} = P(X_{t+n+1} = E_i \mid X_{t+1} = E_{j_1}, X_{t+2} = E_{j_2}, \ldots, X_{t+n} = E_{j_n}\}$$

$(G = (j_1, j_2, \ldots, j_n)^T)$ to be the probability that the demand at time $(t+n+1)$ is i given that the demand at the time $t + k$ is $j_k \in \{1, 2, \ldots, m\}$ for $k = 1, 2, \ldots, n$. Here E_i is an unit vector representing the state of demand. This means that the demand distribution at time $(t + n + 1)$ depends only on the states of the demand at the time $t+1, t+2, \ldots, t+n$, and this is also true for the optimal ordering policy. In the higher-order Markov chain model (3.26), we have

$$p_{i,G} = \sum_{i=1}^{n} \lambda_i Q_i E_{j_i}$$

Under some practical conditions as described in previous sections, one can show that

$$\lim_{t \to \infty} P(X_{t+1} = E_{j_1}, X_{t+2} = E_{j_2}, \ldots, X_{t+n} = E_{j_n}) = s_G$$

where s_G is independent of t. Let

$$r_G, \quad (G = (j_1, j_2, \ldots, j_n)^T)$$

be the ordering policy when the demands of the previous n periods are $j_1, j_2, \ldots, j_n$. The expected cost for all ordering policies $G \in \Phi$ is then given by

$$\mathcal{E}(\Phi) = \sum_{G \in \Phi} s_G \left(\sum_{i=1}^{m} C(r_G, i) p_{i,G} \right). \tag{6.36}$$

The optimal ordering policy $\{r_G^* \mid G \in \Phi\}$ is the one which minimizes (6.36). We remark the computational complexity for computing all the optimal ordering policies r_G^* is of $O(m^n)$ operations because $|\Phi| = m^n$. However, we observe that if the demands of the previous n periods are $j_1, j_2, \ldots, j_n$, then we only need to solve the ordering size r_G which minimizes the expected cost. Since

$$\min_{r_G} \mathcal{E}(\Phi) = \sum_{j=1}^{m} s_G \times \left(\min_{r_G} \sum_{i=1}^{m} C(r_G, i) p_{i,G} \right), \qquad (6.37)$$

the optimal ordering size r_G^* can be obtained by solving

$$\min_{r_G} \sum_{i=1}^{m} C(r_G, i) p_{i,G}, \quad r_G \in \{1, 2, \ldots, m\}.$$

By Proposition 6.6 again, if

$$C(r_G, i) = \begin{cases} C_o(r_G - i) & \text{if } r_G \geq i \\ C_s(i - r_G) & \text{if } r_G < i \end{cases}$$

and let

$$F_G(k) = \sum_{i=1}^{k} p_{i,G}$$

then the optimal ordering size r_G^* satisfies the inequalities

$$F_G(r_G^* - 1) < \frac{C_s}{C_s + C_o} \leq F_G(r_G^*).$$

Therefore, in order to compute the optimal ordering size, the main task here is to estimate the probabilities $p_{i,G}$ or equivalently to estimate the parameters λ_i and Q_i based on the observed data sequence.

6.5.2 A Numerical Example

In this subsection, we present an application of the higher-order Markov model to a generalized Newsboy's problem [57]. The background is that a large soft-drink company faces an in-house problem of production planning and inventory control. There are three types of products A, B and C having five different possible sales volume (1, 2, 3, 4 and 5). Such labelling is useful from both marketing and production planning points of view. The categorical data sequences for the demands of three products of the soft-drink company for some sales periods can be found in [57]. Based on the sales demand data, we build the higher-order Markov models of different orders. These models are then applied to the problem of long-run production planning and the following cost matrix is assumed

$$C = \begin{pmatrix} 0 & 100 & 300 & 700 & 1500 \\ 100 & 0 & 100 & 300 & 700 \\ 300 & 100 & 0 & 100 & 300 \\ 700 & 300 & 100 & 0 & 100 \\ 1500 & 700 & 300 & 100 & 0 \end{pmatrix}.$$

Here $[C]_{ij}$ is the cost when the production plan is for sales volume of state i and the actual sales volume is state j. We note that the costs here are non-linear, i.e. $[C]_{ij} \neq c|i-j|$, where c is a positive constant. When the unsatisfied demand is higher, the shortage cost is larger. Similarly, when the holding product is more, the overage cost is larger. For the higher-order Markov model, we find that the third-order model gives the best optimal cost. Here we also report the results on the first-order model and the stationary model for the three product demand sequences. The results are given in Table 6.5 (taken from [57]).

Table 6.5. The optimal costs of the three different models.

	Product A	Product B	Product C
Third-order Markov Model	11200	9300	10800
First-order Markov Model	27600	18900	11100
Stationary Model	31900	18900	16300

6.6 Summary

In this chapter, a higher-order Markov chain model is proposed with estimation methods for the model parameters. The higher-order Markov chain model is then applied to a number of applications such as DNA sequences, sales demand predictions and web page predictions, Newsboy's problem. Further extension of the model is also discussed.

7

Multivariate Markov Chains

7.1 Introduction

By making use of the transition probability matrix in Chapter 6, a categorical data sequence of m states can be modeled by an m-state Markov chain model. In this chapter, we extend this idea to model multiple categorical data sequences. One would expect categorical data sequences generated by similar sources or same source to be correlated to each other. Therefore by exploring these relationships, one can develop better models for the categorical data sequences and hence better prediction rules.

   The outline of this chapter is as follows. In Section 7.1, we present the multivariate Markov chain model with estimation methods for the model parameters. In Section 7.3, we apply the model to multi-product demand estimation problem. In Section 7.4, an application to credit rating is discussed. In Section 7.5, an application to multiple DNA sequences is presented. In Section 7.6, we apply the model to genetic networks. In Section 7.7, we extend the model to a higher-order multivariate Markov chain model. Finally, a summary is given in Section 7.8 to conclude the chapter.

7.2 Construction of Multivariate Markov Chain Models

In this section, we propose a multivariate Markov chain model to represent the behavior of multiple categorical sequences generated by similar sources or same source. Here we assume that there are s categorical sequences and each has m possible states in the set

$$M = \{1, 2, \ldots, m\}.$$

Let $\mathbf{X}_n^{(j)}$ be the state vector of the jth sequence at time n. If the jth sequence is in state l at time n then we write

$$\mathbf{X}_n^{(j)} = \mathbf{e}_l = (0,\ldots,0,\underbrace{1}_{j\text{th entry}},0\ldots,0)^t.$$

In the proposed multivariate Markov chain model, we assume the following relationship:

$$\mathbf{X}_{n+1}^{(j)} = \sum_{k=1}^{s} \lambda_{jk} P^{(jk)} \mathbf{X}_n^{(k)}, \quad \text{for} \quad j = 1, 2, \ldots, s \tag{7.1}$$

where

$$\lambda_{jk} \geq 0, \quad 1 \leq j, k \leq s \tag{7.2}$$

and

$$\sum_{k=1}^{s} \lambda_{jk} = 1, \quad \text{for} \quad j = 1, 2, \ldots, s. \tag{7.3}$$

The state probability distribution of the kth sequence at time $(n+1)$ depends on the weighted average of $P^{(jk)}\mathbf{X}_n^{(k)}$. Here $P^{(jk)}$ is a transition probability matrix from the states in the kth sequence to the states in the jth sequence, and $\mathbf{X}_n^{(k)}$ is the state probability distribution of the kth sequences at time n. In matrix form we write

$$\mathbf{X}_{n+1} \equiv \begin{pmatrix} \mathbf{X}_{n+1}^{(1)} \\ \mathbf{X}_{n+1}^{(2)} \\ \vdots \\ \mathbf{X}_{n+1}^{(s)} \end{pmatrix} = \begin{pmatrix} \lambda_{11}P^{(11)} & \lambda_{12}P^{(12)} & \cdots & \lambda_{1s}P^{(1s)} \\ \lambda_{21}P^{(21)} & \lambda_{22}P^{(22)} & \cdots & \lambda_{2s}P^{(2s)} \\ \vdots & \vdots & \vdots & \vdots \\ \lambda_{s1}P^{(s1)} & \lambda_{s2}P^{(s2)} & \cdots & \lambda_{ss}P^{(ss)} \end{pmatrix} \begin{pmatrix} \mathbf{X}_n^{(1)} \\ \mathbf{X}_n^{(2)} \\ \vdots \\ \mathbf{X}_n^{(s)} \end{pmatrix}$$

$$\equiv Q\mathbf{X}_n$$

or

$$\mathbf{X}_{n+1} = Q\mathbf{X}_n.$$

Although the column sum of Q is not equal to one (the column sum of $P^{(jk)}$ is equal to one), we still have the following proposition.

Proposition 7.1. *If the parameters $\lambda_{jk} > 0$ for $1 \leq j, k \leq s$, then the matrix Q has an eigenvalue equal to one and the eigenvalues of Q have modulus less than or equal to one.*

Proof. By using (7.2), the column sum of the following matrix

$$\Lambda = \begin{pmatrix} \lambda_{1,1} & \lambda_{2,1} & \cdots & \lambda_{s,1} \\ \lambda_{1,2} & \lambda_{2,2} & \cdots & \lambda_{s,2} \\ \vdots & \vdots & \vdots & \vdots \\ \lambda_{1,s} & \lambda_{2,s} & \cdots & \lambda_{s,s} \end{pmatrix}$$

is equal one. Since $\lambda_{jk} > 0$, Λ is nonnegative and irreducible. By Perron-Frobenius Theorem, there exists a vector

$$\mathbf{y} = (y_1, y_2, \ldots, y_s)^T$$

such that

$$\mathbf{y}^T \Lambda = \mathbf{y}^T.$$

We note that

$$\mathbf{1}_m P^{(ij)} = \mathbf{1}_m, \quad 1 \le i, j \le s,$$

where $\mathbf{1}_m$ is the $1 \times m$ vector of all ones, i.e.,

$$\mathbf{1}_m = (1, 1, \ldots, 1).$$

Then it is easy to show that we have

$$(y_1 \mathbf{1}_m, y_2 \mathbf{1}_m, \ldots, y_s \mathbf{1}_m)Q = (y_1 \mathbf{1}_m, y_2 \mathbf{1}_m, \ldots, y_s \mathbf{1}_m).$$

and hence one must be an eigenvalue of Q.

We then show that all the eigenvalues of Q are less than or equal to one. Let us define the following vector-norm

$$||\mathbf{z}||_V = \max_{1 \le i \le s} \{||\mathbf{z}_i||_1 : \mathbf{z} = (\mathbf{z}_1, \mathbf{z}_2, \cdots, \mathbf{z}_s), \ \mathbf{z}_j \in \mathbf{R}^m, \ 1 \le j \le s\}.$$

It is straightforward to show that $|| \cdot ||_V$ is a vector-norm on $\mathbf{R}^{ms}$. It follows that we can define the following matrix norm

$$||Q||_M \equiv \sup \{||Q\mathbf{z}||_V : ||\mathbf{z}||_V = 1\}.$$

Since $P^{(ij)}$ is a transition matrix, each element of $P^{(ij)}$ are less than or equal to 1. We have

$$||P^{(ij)}\mathbf{z}_j||_1 \le ||\mathbf{z}_j||_1 \le 1, \quad 1 \le i, j \le s.$$

Here $||.||_1$ is the 1-norm for a vector. It follows that

$$||\lambda_{i1} P^{(i1)}\mathbf{z}_1 + \lambda_{i2} P^{(i2)}\mathbf{z}_2 + \cdots + \lambda_{is} P^{(is)}\mathbf{z}_s||_1 \le ||\mathbf{z}||_V \cdot \sum_{j=1}^{s} \lambda_{ij} = 1, \ 1 \le i \le s$$

and hence $||Q||_M \le 1$. Since the spectral radius of Q is always less than or equal to any matrix norm of Q, the result follows.

Proposition 7.2. *Suppose that the matrices $P^{(jk)}$ ($1 \le j, k \le s$) are irreducible and $\lambda_{jk} > 0$ for $1 \le j, k \le s$. Then there is a unique vector*

$$\mathbf{x} = (\mathbf{x}^{(1)}, \mathbf{x}^{(2)}, \ldots, \mathbf{x}^{(s)})^T$$

such that $\mathbf{x} = Q\mathbf{x}$ and

$$\sum_{i=1}^{m} [\mathbf{x}^{(j)}]_i = 1, \ 1 \le j \le s.$$

Proof. By Proposition 7.1, there is exactly one eigenvalue of Q equal to one. This implies that

$$\lim_{n \to \infty} Q^n = \mathbf{v}\mathbf{u}^T$$

is a positive rank one matrix as Q is irreducible. Therefore we have

$$\lim_{n \to \infty} \mathbf{x}_{n+1} = \lim_{n \to \infty} Q\mathbf{x}_n = \lim_{n \to \infty} Q^n \mathbf{x}_0 = \mathbf{v}\mathbf{u}^T \mathbf{x}_0 = \alpha \mathbf{v}.$$

Here α is a positive number since $\mathbf{x} \neq 0$ and is nonnegative. This implies that $\mathbf{x}_n$ tends to a stationary vector as n goes to infinity. Finally, we note that if $\mathbf{x}_0$ is a vector such that

$$\sum_{i=1}^{m} [\mathbf{x}_0^{(j)}]_i = 1, \ 1 \leq j \leq s,$$

then $Q\mathbf{x}_0$ and $\mathbf{x}$ are also vectors having this property.

Now Suppose that there exists $\mathbf{y}$ such that $\mathbf{y} \neq \mathbf{x}$ and

$$\mathbf{y} = \lim_{n \to \infty} \mathbf{x}_n.$$

Then we have

$$||\mathbf{x} - \mathbf{y}|| = ||\mathbf{x} - Q\mathbf{x}|| = 0.$$

This is a contradiction and therefore the vector $\mathbf{x}$ must be unique. Hence the result follows.

We note that $\mathbf{x}$ is not a probability distribution vector, but $\mathbf{x}^{(j)}$ is a probability distribution vector. The above proposition suggests one possible way to estimate the model parameters λ_{ij}. The idea is to find λ_{ij} which minimizes $||Q\hat{\mathbf{x}} - \hat{\mathbf{x}}||$ under certain vector norm $|| \cdot ||$.

7.2.1 Estimations of Model Parameters

In this subsection we propose some methods for the estimations of $P^{(jk)}$ and λ_{jk}. For each data sequence, we estimate the transition probability matrix by the following method. Given the data sequence, we count the transition frequency from the states in the kth sequence to the states in the jth sequence. Hence one can construct the transition frequency matrix for the data sequence. After making a normalization, the estimates of the transition probability matrices can also be obtained. We note that one has to estimate s^2 $m \times m$ transition frequency matrices for the multivariate Markov chain model. More precisely, we count the transition frequency $f_{i_j i_k}^{(jk)}$ from the state i_k in the sequence $\{x_n^{(k)}\}$ to the state i_j in the sequence $\{x_n^{(j)}\}$ and therefore the transition frequency matrix for the sequences can eb constructed as follows:

$$F^{(jk)} = \begin{pmatrix} f_{11}^{(jk)} & \cdots\cdots & f_{m1}^{(jk)} \\ f_{12}^{(jk)} & \cdots\cdots & f_{m2}^{(jk)} \\ \vdots & \vdots & \vdots & \vdots \\ f_{1m}^{(jk)} & \cdots\cdots & f_{mm}^{(jk)} \end{pmatrix}.$$

From $F^{(jk)}$, we get the estimates for $P^{(jk)}$ as follows:

$$\hat{P}^{(jk)} = \begin{pmatrix} \hat{p}_{11}^{(jk)} & \cdots\cdots & \hat{p}_{m1}^{(jk)} \\ \hat{p}_{12}^{(jk)} & \cdots\cdots & \hat{p}_{m2}^{(jk)} \\ \vdots & \vdots & \vdots & \vdots \\ \hat{p}_{1m}^{(jk)} & \cdots\cdots & \hat{p}_{mm}^{(jk)} \end{pmatrix}$$

where

$$\hat{p}_{i_j i_k}^{(jk)} = \begin{cases} \dfrac{f_{i_j i_k}^{(jk)}}{\sum\limits_{i_k=1}^{m} f_{i_j i_k}^{(jk)}} & \text{if } \sum\limits_{i_k=1}^{m} f_{i_j i_k}^{(jk)} \neq 0 \\[1em] 0 & \text{otherwise.} \end{cases}$$

Besides the estimates of $P^{(jk)}$, one needs to estimate the parameters λ_{jk}. We have seen that the multivariate Markov chain model has a stationary vector $\mathbf{x}$ in Proposition 7.2. The vector $\mathbf{x}$ can be estimated from the sequences by computing the proportion of the occurrence of each state in each of the sequences, and let us denote it by

$$\hat{\mathbf{x}} = (\hat{\mathbf{x}}^{(1)}, \hat{\mathbf{x}}^{(2)}, \dots, \hat{\mathbf{x}}^{(s)})^T.$$

One would expect that

$$\begin{pmatrix} \lambda_{11}P^{(11)} & \lambda_{12}P^{(12)} & \cdots & \lambda_{1s}P^{(1s)} \\ \lambda_{21}P^{(21)} & \lambda_{22}P^{(22)} & \cdots & \lambda_{2s}P^{(2s)} \\ \vdots & \vdots & \vdots & \vdots \\ \lambda_{s1}P^{(s1)} & \lambda_{s2}P^{(s2)} & \cdots & \lambda_{ss}P^{(ss)} \end{pmatrix} \hat{\mathbf{x}} \approx \hat{\mathbf{x}}. \tag{7.4}$$

From (7.4), it suggests one possible way to estimate the parameters $\lambda = \{\lambda_{jk}\}$ as follows. In fact, by using $||.||_\infty$ as the vector norm for measuring the difference in (7.4), one may consider solving the following minimization problem:

$$\begin{cases} \min\limits_{\lambda} \max\limits_{i} \left| \left[\sum\limits_{k=1}^{m} \lambda_{jk} \hat{P}^{(jk)} \hat{\mathbf{x}}^{(k)} - \hat{\mathbf{x}}^{(j)} \right]_i \right| \\ \text{subject to} \\ \qquad \sum\limits_{k=1}^{s} \lambda_{jk} = 1, \\ \text{and} \\ \qquad \lambda_{jk} \geq 0, \quad \forall k. \end{cases} \tag{7.5}$$

Problem (7.5) can be formulated as s linear programming problems as follows, see for instance [79].

For each j:

$$\begin{cases} \min\limits_{\lambda} w_j \\ \text{subject to} \\ \qquad \begin{pmatrix} w_j \\ w_j \\ \vdots \\ w_j \end{pmatrix} \geq \hat{\mathbf{x}}^{(j)} - B \begin{pmatrix} \lambda_{j1} \\ \lambda_{j2} \\ \vdots \\ \lambda_{js} \end{pmatrix}, \\ \qquad \begin{pmatrix} w_j \\ w_j \\ \vdots \\ w_j \end{pmatrix} \geq -\hat{\mathbf{x}}^{(j)} + B \begin{pmatrix} \lambda_{j1} \\ \lambda_{j2} \\ \vdots \\ \lambda_{js} \end{pmatrix}, \\ \qquad w_j \geq 0, \\ \qquad \sum\limits_{k=1}^{s} \lambda_{jk} = 1, \quad \lambda_{jk} \geq 0, \quad \forall k, \end{cases}$$

where

$$B = [\hat{P}^{(j1)} \hat{\mathbf{x}}^{(1)} \mid \hat{P}^{(j2)} \hat{\mathbf{x}}^{(2)} \mid \cdots \mid \hat{P}^{(js)} \hat{\mathbf{x}}^{(s)}].$$

In the next subsection, we give an example to demonstrate the construction of a multivariate Markov chain model from two data sequences.

7.2.2 An Example

Consider the following two categorical data sequences:

$$S_1 = \{4, 3, 1, 3, 4, 4, 3, 3, 1, 2, 3, 4\}$$

and

$$S_2 = \{1, 2, 3, 4, 1, 4, 4, 3, 3, 1, 3, 1\}.$$

By counting the transition frequencies

$$S_1 : 4 \to 3 \to 1 \to 3 \to 4 \to 4 \to 3 \to 3 \to 1 \to 2 \to 3 \to 4$$

and

$$S_2 : 1 \to 2 \to 3 \to 4 \to 1 \to 4 \to 4 \to 3 \to 3 \to 1 \to 3 \to 1$$

we have

$$F^{(11)} = \begin{pmatrix} 0\,0\,2\,0 \\ 1\,0\,0\,0 \\ 1\,1\,1\,2 \\ 0\,0\,2\,1 \end{pmatrix} \quad \text{and} \quad F^{(22)} = \begin{pmatrix} 0\,0\,2\,1 \\ 1\,0\,0\,0 \\ 1\,1\,1\,1 \\ 1\,0\,1\,1 \end{pmatrix}.$$

Moreover by counting the inter-transition frequencies

$$
\begin{array}{ccccccccccccc}
S_1 : 4 & & 3 & & 1 & & 3 & & 4 & & 4 & & 3 & & 3 & & 1 & & 2 & & 3 & & 4 \\
 & \nearrow & & \nearrow & & \nearrow & & \nearrow & & \nearrow & & \nearrow & & \nearrow & & \nearrow & & \nearrow & & \nearrow & & \nearrow & \\
S_2 : 1 & & 2 & & 3 & & 4 & & 1 & & 4 & & 4 & & 3 & & 3 & & 1 & & 3 & & 1
\end{array}
$$

and

$$
\begin{array}{ccccccccccccc}
S_1 : 4 & & 3 & & 1 & & 3 & & 4 & & 4 & & 3 & & 3 & & 1 & & 2 & & 3 & & 4 \\
 & \searrow & & \searrow & & \searrow & & \searrow & & \searrow & & \searrow & & \searrow & & \searrow & & \searrow & & \searrow & & \searrow & \\
S_2 : 1 & & 2 & & 3 & & 4 & & 1 & & 4 & & 4 & & 3 & & 3 & & 1 & & 3 & & 1
\end{array}
$$

we have

$$F^{(21)} = \begin{pmatrix} 1\,0\,2\,0 \\ 0\,0\,0\,1 \\ 0\,1\,3\,0 \\ 1\,0\,0\,2 \end{pmatrix}, \qquad F^{(12)} = \begin{pmatrix} 0\,1\,1\,0 \\ 0\,0\,1\,0 \\ 2\,0\,1\,2 \\ 1\,0\,1\,1 \end{pmatrix}.$$

After making a normalization, we have the transition probability matrices:

$$\hat{P}^{(11)} = \begin{pmatrix} 0 & 0 & \frac{2}{5} & 0 \\ \frac{1}{2} & 0 & 0 & 0 \\ \frac{1}{2} & 1 & \frac{1}{5} & \frac{2}{3} \\ 0 & 0 & \frac{2}{5} & \frac{1}{3} \end{pmatrix}, \qquad \hat{P}^{(12)} = \begin{pmatrix} 0 & 1 & \frac{1}{4} & 0 \\ 0 & 0 & \frac{1}{4} & 0 \\ \frac{2}{3} & 0 & \frac{1}{4} & \frac{2}{3} \\ \frac{1}{3} & 0 & \frac{1}{4} & \frac{1}{3} \end{pmatrix},$$

$$\hat{P}^{(21)} = \begin{pmatrix} \frac{1}{2} & 0 & \frac{2}{5} & 0 \\ 0 & 0 & 0 & \frac{1}{3} \\ 0 & 1 & \frac{3}{5} & 0 \\ \frac{1}{2} & 0 & 0 & \frac{2}{3} \end{pmatrix}, \qquad \hat{P}^{(22)} = \begin{pmatrix} 0 & 0 & \frac{1}{2} & \frac{1}{3} \\ \frac{1}{3} & 0 & 0 & 0 \\ \frac{1}{3} & 1 & \frac{1}{4} & \frac{1}{3} \\ \frac{1}{3} & 0 & \frac{1}{4} & \frac{1}{3} \end{pmatrix}.$$

Moreover we also have

$$\hat{\mathbf{x}}_1 = (\frac{1}{6}, \frac{1}{12}, \frac{5}{12}, \frac{1}{3})^T \quad \text{and} \quad \hat{\mathbf{x}}_2 = (\frac{1}{3}, \frac{1}{12}, \frac{1}{3}, \frac{1}{4})^T.$$

By solving the corresponding linear programming problems, the multivariate Markov chain models for the two categorical data sequences S_1 and S_2 are then given by

$$\begin{cases} \mathbf{x}_{n+1}^{(1)} = 0.5000 \hat{P}^{(11)} \mathbf{x}_n^{(1)} + 0.5000 \hat{P}^{(12)} \mathbf{x}_n^{(2)} \\ \mathbf{x}_{n+1}^{(2)} = 0.8858 \hat{P}^{(21)} \mathbf{x}_n^{(1)} + 0.1142 \hat{P}^{(22)} \mathbf{x}_n^{(2)}. \end{cases}$$

7.3 Applications to Multi-product Demand Estimation

Let us consider demand estimation problems stated as in Section 6.3.2. We study the customer's sales demand of five important products of the company in a year. The sales demand sequences are generated by the same customer and therefore we expect that they should be correlated to each other. Therefore by exploring these relationships, one can develop the multivariate Markov chain model for such demand sequences, hence obtain better prediction rules.

We first estimate all the transition probability matrices $P^{(ij)}$ by using the method proposed in Section 7.2 and we also have the estimates of the state distribution of the five products:

$$\begin{cases} \hat{\mathbf{x}}_1 = (0.0818, 0.4052, 0.0483, 0.0335, 0.0037, 0.4275)^T, \\ \hat{\mathbf{x}}_2 = (0.3680, 0.1970, 0.0335, 0.0000, 0.0037, 0.3978)^T, \\ \hat{\mathbf{x}}_3 = (0.1450, 0.2045, 0.0186, 0.0000, 0.0037, 0.6283)^T, \\ \hat{\mathbf{x}}_4 = (0.0000, 0.3569, 0.1338, 0.1896, 0.0632, 0.2565)^T, \\ \hat{\mathbf{x}}_5 = (0.0000, 0.3569, 0.1227, 0.2268, 0.0520, 0.2416)^T. \end{cases}$$

By solving the corresponding minimization problems through linear programming we obtain the optimal solution:

$$\Lambda = [\lambda_{jk}] = \begin{pmatrix} 0.0000 & 1.0000 & 0.0000 & 0.0000 & 0.0000 \\ 0.0000 & 1.0000 & 0.0000 & 0.0000 & 0.0000 \\ 0.0000 & 0.0000 & 0.0000 & 0.0000 & 1.0000 \\ 0.0000 & 0.0000 & 0.0000 & 0.4741 & 0.5259 \\ 0.0000 & 0.0000 & 0.0000 & 1.0000 & 0.0000 \end{pmatrix}$$

and the multivariate Markov chain model for these five sequences is as follows:

$$\begin{cases} \mathbf{x}_{n+1}^{(1)} = P^{(12)}\mathbf{x}_n^{(2)} \\ \mathbf{x}_{n+1}^{(2)} = P^{(22)}\mathbf{x}_n^{(2)} \\ \mathbf{x}_{n+1}^{(3)} = P^{(35)}\mathbf{x}_n^{(5)} \\ \mathbf{x}_{n+1}^{(4)} = 0.4741 P^{(44)}\mathbf{x}_n^{(4)} + 0.5259 P^{(45)}\mathbf{x}_n^{(5)} \\ \mathbf{x}_{n+1}^{(5)} = P^{(54)}\mathbf{x}_n^{(4)} \end{cases}$$

where

$$P^{(12)} = \begin{pmatrix} 0.0707 & 0.1509 & 0.0000 & 0.2000 & 0.0000 & 0.0660 \\ 0.4343 & 0.4528 & 0.4444 & 0.2000 & 1.0000 & 0.3491 \\ 0.0101 & 0.1321 & 0.2222 & 0.2000 & 0.0000 & 0.0283 \\ 0.0101 & 0.0943 & 0.2222 & 0.2000 & 0.0000 & 0.0094 \\ 0.0000 & 0.0000 & 0.2000 & 0.0000 & 0.0000 & 0.0094 \\ 0.4747 & 0.1698 & 0.1111 & 0.2000 & 0.0000 & 0.5377 \end{pmatrix}$$

$$P^{(22)} = \begin{pmatrix} 0.4040 & 0.2075 & 0.0000 & 0.2000 & 1.0000 & 0.4340 \\ 0.1111 & 0.4717 & 0.3333 & 0.2000 & 0.0000 & 0.1321 \\ 0.0202 & 0.0566 & 0.3333 & 0.2000 & 0.0000 & 0.0094 \\ 0.0000 & 0.0000 & 0.0000 & 0.2000 & 0.0000 & 0.0000 \\ 0.0000 & 0.0000 & 0.1111 & 0.2000 & 0.0000 & 0.0000 \\ 0.4646 & 0.2642 & 0.2222 & 0.2000 & 0.0000 & 0.4245 \end{pmatrix}$$

$$P^{(35)} = \begin{pmatrix} 0.2000 & 0.0947 & 0.1515 & 0.1639 & 0.0714 & 0.2154 \\ 0.2000 & 0.1895 & 0.2727 & 0.2295 & 0.1429 & 0.1846 \\ 0.2000 & 0.0421 & 0.0000 & 0.0000 & 0.0000 & 0.0154 \\ 0.2000 & 0.0000 & 0.0000 & 0.0000 & 0.0000 & 0.0000 \\ 0.2000 & 0.0105 & 0.0000 & 0.0000 & 0.0000 & 0.0000 \\ 0.2000 & 0.6632 & 0.5758 & 0.6066 & 0.7857 & 0.5846 \end{pmatrix}$$

$$P^{(44)} = \begin{pmatrix} 0.2000 & 0.0000 & 0.0000 & 0.0000 & 0.0000 & 0.0000 \\ 0.2000 & 0.4947 & 0.1389 & 0.0196 & 0.0588 & 0.6087 \\ 0.2000 & 0.0842 & 0.3056 & 0.1765 & 0.0588 & 0.1014 \\ 0.2000 & 0.0000 & 0.3056 & 0.5686 & 0.5294 & 0.0290 \\ 0.2000 & 0.0105 & 0.0556 & 0.1569 & 0.3529 & 0.0000 \\ 0.2000 & 0.4105 & 0.1944 & 0.0784 & 0.0000 & 0.2609 \end{pmatrix}$$

$$P^{(45)} = \begin{pmatrix} 0.2000 & 0.0000 & 0.0000 & 0.0000 & 0.0000 & 0.0000 \\ 0.2000 & 0.4737 & 0.2121 & 0.0328 & 0.0000 & 0.6462 \\ 0.2000 & 0.1053 & 0.2121 & 0.1967 & 0.0714 & 0.0923 \\ 0.2000 & 0.0000 & 0.2424 & 0.5410 & 0.5714 & 0.0308 \\ 0.2000 & 0.0105 & 0.0303 & 0.1803 & 0.2857 & 0.0000 \\ 0.2000 & 0.4105 & 0.3030 & 0.0492 & 0.0714 & 0.2308 \end{pmatrix}$$

$$P^{(54)} = \begin{pmatrix} 0.2000 & 0.0000 & 0.0000 & 0.0000 & 0.0000 & 0.0000 \\ 0.2000 & 0.4842 & 0.1667 & 0.0196 & 0.0588 & 0.6087 \\ 0.2000 & 0.1053 & 0.1667 & 0.1569 & 0.0588 & 0.1159 \\ 0.2000 & 0.0000 & 0.4444 & 0.6275 & 0.6471 & 0.0290 \\ 0.2000 & 0.0105 & 0.0278 & 0.1569 & 0.2353 & 0.0000 \\ 0.2000 & 0.4000 & 0.1944 & 0.0392 & 0.0000 & 0.2464 \end{pmatrix}.$$

According to the multivariate Markov chain model, Products A and B are closely related. In particular, the sales demand of Product A depends strongly on Product B. The main reason is that the chemical nature of Products A and B is the same, but they have different packaging for marketing purposes. Moreover, Products C, D and E are closely related. Similarly, products C and E have the same product flavor, but different packaging. It is interesting to note that even through Products D and E have different chemical nature but similar flavor, the results show that their sales demand are also closely related.

Next we use the multivariate Markov chain model, to make predictions on the state $\hat{\mathbf{x}}_t$ at time t which can be taken as the state with the maximum probability, i.e.,

$$\hat{\mathbf{x}}_t = j, \quad \text{if } [\hat{\mathbf{x}}_t]_i \leq [\hat{\mathbf{x}}_t]_j, \forall 1 \leq i \leq m.$$

To evaluate the performance and effectiveness of our multivariate Markov chain model, a prediction result is measured by the prediction accuracy r defined as

$$r = \frac{1}{T} \times \sum_{t=n+1}^{T} \delta_t \times 100\%,$$

where T is the length of the data sequence and

$$\delta_t = \begin{cases} 1, & \text{if } \hat{\mathbf{x}}_t = \mathbf{x}_t \\ 0, & \text{otherwise.} \end{cases}$$

For the sake of comparison, we also give the results for the first-order Markov chain model of individual sales demand sequence. The results are reported in Table 7.1. There is noticeable improvement in prediction accuracy in Product A while improvements are also observed in Product D and Product E. The results show the effectiveness of our multivariate Markov chain model.

Table 7.1. Prediction accuracy in the sales demand data.

	Product A	Product B	Product C	Product D	Product E
First-order Markov Chain	46%	45%	63%	51%	53%
Multivariate Markov Chain	50%	45%	63%	52%	55%

7.4 Applications to Credit Rating

In the last decade, there has been a considerable interest in modelling the dependency of the credit risks due to the practical importance and relevance of risk analysis of credit portfolios [6, 7, 20, 30, 85, 86, 87, 88, 90, 93, 120, 119, 122, 161, 164, 168, 182, 210, 211]. The specification of the model that explains and describes the dependency of the credit risks can have significant implications in pricing credit risky securities and managing credit risky portfolios. The discrete-time homogeneous Markov Chain model has been used among academic researchers and market practitioners in modelling the transitions of the ratings of a credit risk over time. The credit transition probability matrix represents the likelihood of the future evolution of the ratings. The credit transition probability matrix can be estimated based on the available empirical data for credit ratings. Standard & Poor and Moodys are the major providers of the credit rating data. They provide and update from time to time the historical data for various individual companies and countries.

Credibility theory has been widely applied in the actuarial discipline for calculating a policyholder's premium through experience rating of the policyholder's past claims. Mowbray [155], Bühlmann [37] and Klugman, Panjer and Willmot [133] provided an excellent account on actuarial credibility theory. Siu and Yang [190] and Siu, Tong and Yang [191] provided some discussions on the use of Bayesian credibility theory for risk measurement. By employing the idea of credibility theory, one can provide an estimate for the credit transition probability matrix as a linear combination of the empirical credit transition probability matrix and a prior credit transition probability matrix [113] et al. Here we consider an approach that can provide an analytically tractable way to estimate credit transition probability matrix. The estimator for transition probability matrices of ratings is a linear combination of a prior matrix given by the empirical transition matrix estimated directly from Standard & Poor's data and a model-based updating matrix evaluated from the ordered probit model. This approach provides market practitioners with an intuitively appealing and convenient way for the estimation of the unknown parameters and credit transition probability matrices in the multivariate Markov chain model Kijima et al [128].

7.4.1 The Credit Transition Matrix

In this subsection, we assume that the estimate of each credit transition probability matrix can be represented as a linear combination of prior credit transition probability matrix and the empirical credit transition probability matrix, where the empirical credit transition probability matrix is calculated based on the transition frequencies of ratings (see Section 7.3). Then, by Proposition 7.1, there exists a vector X of stationary probability distributions, we can estimate the necessary parameters based on the stationary distributions for the ratings.

Let $Q^{(jk)}$ denote the prior credit transition probability matrix. The empirical estimate $\hat{P}^{(jk)}$ of the credit transition probability matrix can be obtained using the method in Section 7.2.1. Here, we specify the prior credit transition probability matrix by the credit transition probability matrix created by Standard & Poor's. The credit transition probability matrix produced by Standard & Poor's has widely been used as a benchmark for credit risk measurement and management in the finance and banking industries. For the purpose of illustration, we assign a common prior credit transition probability matrix for the two credit risky assets as the credit transition probability matrix created by Standard & Poor's to represent the belief that the credit transition probability matrices for the two credit risky assets are essentially the same based on the prior information. If more prior information about the credit rating of each credit risky asset is available, we can determine a more informative prior credit transition probability matrix for each credit risky asset. For a comprehensive overview and detailed discussion on the choice of prior distributions based on prior information, refer to some representative monographs in Bayesian Statistics, such as Lee [139], Bernardo and Smith [17] and Robert [178], etc. Then, the estimate $P_e^{(jk)}$ of the credit transition probability $P^{(jk)}$ is given by

$$P_e^{(jk)} = w_{jk}Q^{(jk)} + (1 - w_{jk})\hat{P}^{(jk)} \ , \ j, k = 1, 2, \ldots, n \ , \qquad (7.6)$$

where $0 \leq w_{jk} \leq 1$, for each $j, k = 1, 2, \ldots, n$. From proposition 7.1, we have that

$$\begin{pmatrix} \lambda_{11}P_e^{(11)} & \lambda_{12}P_e^{(12)} & \cdots & \lambda_{1n}P_e^{(1n)} \\ \lambda_{21}P_e^{(21)} & \lambda_{22}P_e^{(22)} & \cdots & \lambda_{2n}P_e^{(2n)} \\ \vdots & \vdots & \vdots & \vdots \\ \lambda_{n1}P_e^{(n1)} & \lambda_{n2}P_e^{(n2)} & \cdots & \lambda_{nn}P_e^{(nn)} \end{pmatrix} \hat{\mathbf{x}} \approx \hat{\mathbf{x}}. \tag{7.7}$$

Let

$$\tilde{\lambda}_{jk}^1 = \lambda_{jk} w_{jk}$$

and

$$\tilde{\lambda}_{jk}^2 = \lambda_{jk}(1 - w_{jk}).$$

Then, it is easy to check that for each $j, k = 1, 2, \ldots, n$, we have

$$\tilde{\lambda}_{jk}^1 + \tilde{\lambda}_{jk}^2 = \lambda_{jk}.$$

We note that the estimation of λ_{jk} and w_{jk} is equivalent to the estimation of $\tilde{\lambda}_{jk}^1$ and $\tilde{\lambda}_{jk}^2$. Then, (7.7) can be written in the following form:

$$\begin{pmatrix} \tilde{\lambda}_{11}^1 Q^{(11)} + \tilde{\lambda}_{11}^2 \hat{P}^{(11)} & \cdots & \tilde{\lambda}_{1n}^1 Q^{(1n)} + \tilde{\lambda}_{1n}^2 \hat{P}^{(1n)} \\ \tilde{\lambda}_{21}^1 Q^{(21)} + \tilde{\lambda}_{21}^2 \hat{P}^{(21)} & \cdots & \tilde{\lambda}_{2n}^1 Q^{(2n)} + \tilde{\lambda}_{2n}^2 \hat{P}^{(2n)} \\ \vdots & \vdots & \vdots \\ \tilde{\lambda}_{n1}^1 Q^{(n1)} + \tilde{\lambda}_{n1}^2 \hat{P}^{(n1)} & \cdots & \tilde{\lambda}_{nn}^1 Q^{(nn)} + \tilde{\lambda}_{nn}^2 \hat{P}^{(nn)} \end{pmatrix} \hat{\mathbf{X}} \approx \hat{\mathbf{X}}. \tag{7.8}$$

Now, we can formulate our estimation problem as follows:

$$\begin{cases} \min_{\tilde{\lambda}^1, \tilde{\lambda}^2} \max_i \left| \left[\sum_{k=1}^m (\tilde{\lambda}_{jk}^1 Q^{(jk)} + \tilde{\lambda}_{jk}^2 \hat{P}^{(jk)}) \hat{X}^{(k)} - \hat{X}^{(j)} \right]_i \right| \\ \text{subject to} \\ \qquad \sum_{k=1}^n (\tilde{\lambda}_{jk}^1 + \tilde{\lambda}_{jk}^2) = 1, \quad \tilde{\lambda}_{jk}^1 \geq 0 \\ \text{and} \\ \qquad \tilde{\lambda}_{jk}^2 \geq 0, \quad \forall j, k. \end{cases} \tag{7.9}$$

Let

$$O_j = \max_i \left| \left[\sum_{k=1}^m (\tilde{\lambda}_{jk}^1 Q^{(jk)} + \tilde{\lambda}_{jk}^2 \hat{P}^{(jk)}) \hat{\mathbf{x}}^{(k)} - \hat{\mathbf{x}}^{(j)} \right]_i \right|.$$

Then, Problem (7.9) can be re-formulated as the following set of n linear programming problems as in Chapter 6. It is clear that, one can also choose vector $||.||_1$ instead of the vector norm $||.||_\infty$. The resulting problem can be still as a linear programming problem. A detailed application in credit rating can be found in Siu et al. [188].

7.5 Applications to DNA Sequences Modeling

In this section, we test multivariate Markov chain models for DNA sequences and analyze their correlations, Ching et al [66]. Because of its extraordinary position as a preferred model in biochemical genetics, molecular biology, and biotechnology, *Escherichia coli* K-12 was the earliest organism to be suggested as a candidate for whole genome sequencing. The complete genome sequence of *E. coli* was obtained in 1997 [24]. A complete listing of *E. coli* open reading frames (ORFs), that is, long contiguous reading frame without STOP codons, is now available at the website [227]. In the tests, we used this database in all of our computations. The lengths of the DNA sequences we tested are from 1000 to 4000.

In the first test, we tried to use (A, C, G, T) as the set of possible states that a multivariate Markov chain model can take. However, we find that we cannot construct any useful models. Each DNA sequence is independent of the other DNA sequences, i.e., $\lambda_{ii} = 1$ and $\lambda_{ij} = 0$ for $i \neq j$. It is well-known that amino acids are encoded by consecutive sequences of 3 nucleotides, called *codon*. Taking this fact into account, in the construction of multivariate Markov chain model, one identifies 12 symbols: the four nucleotides (A, T, G, C) in the first position, the four letters

$$(A', T', G', C')$$

in the second position and the four same letters

$$(A'', T'', G'', C'')$$

in the third position of a reading frame of period three. Using this approach, alphabet sequence

$$ACTGTT\ldots\ldots$$

is re-written as

$$AC'T''GT'T''\ldots\ldots,$$

and therefore the transition probability for a letter doublet being different according to the position in the hypothetical codon. For instance, below is the transition matrix for the DNA sequence (b2647) in the database:

$$
\left(
\begin{array}{cccc|cccc|cccc}
0 & 0 & 0 & 0 & 0 & 0 & 0 & 0 & 0.4067 & 0.3898 & 0.3109 & 0.3320 \\
0 & 0 & 0 & 0 & 0 & 0 & 0 & 0 & 0.1498 & 0.1332 & 0.1965 & 0.1066 \\
0 & 0 & 0 & 0 & 0 & 0 & 0 & 0 & 0.3303 & 0.3608 & 0.3812 & 0.4344 \\
0 & 0 & 0 & 0 & 0 & 0 & 0 & 0 & 0.1131 & 0.1162 & 0.1114 & 0.1270 \\
\hline
0.3648 & 0.3722 & 0.2400 & 0.2324 & 0 & 0 & 0 & 0 & 0 & 0 & 0 & 0 \\
0.3007 & 0.1570 & 0.2083 & 0.3622 & 0 & 0 & 0 & 0 & 0 & 0 & 0 & 0 \\
0.1352 & 0.1614 & 0.3550 & 0.0865 & 0 & 0 & 0 & 0 & 0 & 0 & 0 & 0 \\
0.1993 & 0.3094 & 0.1967 & 0.3189 & 0 & 0 & 0 & 0 & 0 & 0 & 0 & 0 \\
\hline
0 & 0 & 0 & 0 & 0.2189 & 0.3030 & 0.1173 & 0.1788 & 0 & 0 & 0 & 0 \\
0 & 0 & 0 & 0 & 0.2274 & 0.2576 & 0.3548 & 0.2291 & 0 & 0 & 0 & 0 \\
0 & 0 & 0 & 0 & 0.1684 & 0.2449 & 0.1848 & 0.2821 & 0 & 0 & 0 & 0 \\
0 & 0 & 0 & 0 & 0.3853 & 0.1944 & 0.3431 & 0.3101 & 0 & 0 & 0 & 0 \\
\end{array}
\right).
$$

Because we order the states as

$$(ATGCA'T'G'C'A''T''G''C''),$$

the transition matrix is a 3-by-3 cyclic matrix. The cyclic matrix has nonzero blocks at $(2,1)$th, $(3,2)$th and $(1,3)$th blocks and other blocks are zero. This structure allows us to implement the multivariate Markov chain model more efficiently in the estimation of the parameters.

E. coli has been a paradigm for the identification of motifs. The basic idea for identifying significant motifs is to design, a priori, a probabilistic model permitting generation of a theoretical genetic sequence and then compute the expected frequency of a given motif in this model-derived sequence. This latter theoretical motif frequency is subsequently compared with the frequency observed in the real sequence. If the difference between the two frequencies is important, one can surmise that the *motif* reflects a process of biological significance (c.f. [108]). Several periodic Markov chain models have been introduced for this purpose, see for instance [28] and [131]. Our model is different from the previous ones in the sense that we used the information from more than one ORF sequences. This approach may be useful if certain 'style' exists within the genes of the organism (in fact, codon usage biases do exist in *E. coli*).

We have tried to construct the multivariate Markov chain models for the DNA sequences in the database of *E. coli*. Some results for modeling DNA sequences are reported in Table 7.2. In Table 7.2, the target DNA sequences in the first column means that the multivariate Markov chain models are constructed for these DNA sequences. The DNA sequences in the second column are the related DNA sequences in the multivariate Markov chain model for the target DNA sequence. The number in the bracket is the weighting parameter (λ_{jk}) of the related DNA sequence in the multivariate Markov chain model. For instance, the model for the DNA sequence (b0890) is as follows:

$$\mathbf{X}_n^{(b0890)} = 0.918\hat{P}^{(b0890\ b3593)}\mathbf{X}_n^{b3593} + 0.082\hat{P}^{(b0890\ b0890)}\mathbf{X}_n^{(b0890)}.$$

We see from Table 7.2 that there are some DNA sequences depending only on the other DNA sequences, e.g.,

$$b4289, b2150, b1320, b4232, b2411, b2645,$$

and

$$b0344, b1687, b3894, b1510, b1014, b2557.$$

These DNA sequences were selected to evaluate their biological functions and understand their dependence of other DNA sequences.

We would like to consider the state vector $\mathbf{X}_n^{(b0924)}$ of the DNA sequence (b0924) at the base n depends on the state vectors $\mathbf{X}_n^{(b2647)}$ of the DNA sequence (b2647), and itself. More precisely, we have the following multivariate Markov chain model:

$$\mathbf{X}_n^{(b0924)} = 0.356\hat{P}^{(b0924\ b2647)}\mathbf{X}_n^{b2647} + 0.644\hat{P}^{(b0924\ b0924)}\mathbf{X}_n^{(b0924)}.$$

The transition matrices $\hat{P}^{(b0924\ b2647)}$ and $\hat{P}^{(b0924\ b0924)}$ are given by

$$
\begin{pmatrix}
0 & 0 & 0 & 0 & 0 & 0 & 0 & 0 & 0.1465 & 0.1853 & 0.2197 & 0.2263 \\
0 & 0 & 0 & 0 & 0 & 0 & 0 & 0 & 0.3248 & 0.3553 & 0.2962 & 0.3060 \\
0 & 0 & 0 & 0 & 0 & 0 & 0 & 0 & 0.4108 & 0.3198 & 0.3662 & 0.3621 \\
0 & 0 & 0 & 0 & 0 & 0 & 0 & 0 & 0.1178 & 0.1396 & 0.1178 & 0.1056 \\
0.3556 & 0.3146 & 0.3763 & 0.3631 & 0 & 0 & 0 & 0 & 0 & 0 & 0 & 0 \\
0.1907 & 0.2347 & 0.1820 & 0.2083 & 0 & 0 & 0 & 0 & 0 & 0 & 0 & 0 \\
0.1796 & 0.2066 & 0.1714 & 0.1548 & 0 & 0 & 0 & 0 & 0 & 0 & 0 & 0 \\
0.2741 & 0.2441 & 0.2703 & 0.2738 & 0 & 0 & 0 & 0 & 0 & 0 & 0 & 0 \\
0 & 0 & 0 & 0 & 0.1530 & 0.1257 & 0.1640 & 0.1751 & 0 & 0 & 0 & 0 \\
0 & 0 & 0 & 0 & 0.2616 & 0.3115 & 0.2397 & 0.2404 & 0 & 0 & 0 & 0 \\
0 & 0 & 0 & 0 & 0.3548 & 0.3403 & 0.3975 & 0.3056 & 0 & 0 & 0 & 0 \\
0 & 0 & 0 & 0 & 0.2306 & 0.2225 & 0.1987 & 0.2789 & 0 & 0 & 0 & 0
\end{pmatrix}
$$

and

$$
\begin{pmatrix}
0 & 0 & 0 & 0 & 0 & 0 & 0 & 0 & 0.2026 & 0.2360 & 0.1618 & 0.2023 \\
0 & 0 & 0 & 0 & 0 & 0 & 0 & 0 & 0.3216 & 0.2335 & 0.3950 & 0.3092 \\
0 & 0 & 0 & 0 & 0 & 0 & 0 & 0 & 0.4009 & 0.3985 & 0.3256 & 0.3497 \\
0 & 0 & 0 & 0 & 0 & 0 & 0 & 0 & 0.0749 & 0.1320 & 0.1175 & 0.1387 \\
0.3605 & 0.3061 & 0.4628 & 0.1798 & 0 & 0 & 0 & 0 & 0 & 0 & 0 & 0 \\
0.1905 & 0.0713 & 0.2695 & 0.3146 & 0 & 0 & 0 & 0 & 0 & 0 & 0 & 0 \\
0.1429 & 0.3040 & 0.1097 & 0.1011 & 0 & 0 & 0 & 0 & 0 & 0 & 0 & 0 \\
0.3061 & 0.3187 & 0.1580 & 0.4045 & 0 & 0 & 0 & 0 & 0 & 0 & 0 & 0 \\
0 & 0 & 0 & 0 & 0.3133 & 0.1065 & 0.0379 & 0.0501 & 0 & 0 & 0 & 0 \\
0 & 0 & 0 & 0 & 0.2026 & 0.2715 & 0.4545 & 0.2180 & 0 & 0 & 0 & 0 \\
0 & 0 & 0 & 0 & 0.2946 & 0.4570 & 0.0720 & 0.5263 & 0 & 0 & 0 & 0 \\
0 & 0 & 0 & 0 & 0.1895 & 0.1649 & 0.4356 & 0.2055 & 0 & 0 & 0 & 0
\end{pmatrix}
$$

respectively. We see that $\hat{P}^{(b0924\ b2647)}$ and $\hat{P}^{(b0924\ b0924)}$ are cyclic matrices. It is interesting to note from our analysis that the DNA sequence ($b2647$) plays an important role in the construction of multivariate Markov chain models of other DNA sequences. We check that this DNA sequence corresponds to outer membrane proteins involved in the so-called *antigenic variation* phenomenon, that allows the cell to escape the immune response of the host.

We also compare the multivariate Markov chain model with the Markov model of a single DNA sequence. The improvement in accuracy of using the multivariate Markov chain model over the Markov chain model of a single DNA sequence is reported in the last column of Table 7.2. We find that the prediction accuracy of using the multivariate Markov chain model is significantly higher than that of using the Markov chain model of a single DNA sequence.

On the other hand, one would like to construct the conventional first-order Markov chain describing multiple DNA sequences. However, such model require a large number of training data (i.e., the length of the DNA sequence should be long enough) to accurately estimate the transition probabilities of each base occurring after every possible combination of the proceeding bases. In the tests, the lengths of short DNA sequences are about 1000 and there are 97% transition probabilities of the conventional model that cannot be estimated. For the long DNA sequences (their lengths are about 4000), there are still 96% transition probabilities of the model that cannot be estimated. Therefore, the applicability of such conventional model is difficult.

Table 7.2. Results of the multivariate Markov chain models.

Target DNA sequences	DNA sequences in the multivariate Markov chain model (weighting parameters)	Improvement in accuracy (%)
b4289	b1415 (1)	56.25
b2150	b3830 (1)	49.00
b2410	b3830 (1)	47.16
b1320	b2410 (0.9963), b2546 (0.0037)	41.32
b4232	b1415 (0.9992), b3830 (0.0008)	36.57
b779	b779 (0.457), b3081 (0.260), b2411 (0.106), b1645 (0.177)	57.81
b3081	b3081 (0.426), b2411 (0.574)	43.02
b1023	b1023 (0.252), b2411 (0.748)	15.40
b2411	b779 (0.476), b1645 (0.524)	39.37
b2645	b1645 (1)	40.70
b1435	b3081 (0.5), b1435 (0.5)	49.09
b2076	b2076 (0.417), b0344 (0.583)	27.83
b0344	b2076 (0.826), b1474 (0.174)	60.07
b1687	b2076 (0.937), b0059 (0.0626)	13.94
b3894	b0344 (1)	27.79
b3593	b3482 (0.453), b3593 (0.547)	36.23
b3987	b3988 (0.081), b0700 (0.668), b3987 (0.171), b1014 (0.080)	54.06
b0890	b3593 (0.818), b0890 (0.182)	30.37
b1510	b3593 (0.685), b3987 (0.315)	37.61
b1014	b3988 (1)	44.43
b2557	b3482 (0.114), b3987 (0.886)	39.23
b0924	b2647 (0.918), b0924 (0.082)	54.53

The advantage of the Markov chain model in biological applications is its effectiveness in prediction. However, its use is limited to a single DNA sequence. The multivariate Markov chain model presented here has removed this limitation whilst preserving its effectiveness. The extension allows us to model multiple DNA sequences directly and analyze them as a whole. Because biological applications deal with a very large number of DNA sequences, scalability is a basic requirement to these applications. Our experimental results have demonstrated that the multivariate Markov chain model is indeed scalable to very large DNA sequences.

7.6 Applications to Genetic Networks

In this section, we applied the multivariate Markov chain model to model genetic networks, Ching et al. [64]. One of the important focus of genomic research is to understand the mechanism in which cells execute and control the huge number of operations for normal functions, and also the way in which the cellular systems fail in disease. Models based on methods such as neural networks, non-linear ordinary, Petri nets, differential equations have been proposed for such problem, see for instance Smolen *et al.* [192], Bower [29] and DeJong [83].

Another approach is to model the genetic regulatory system by a Boolean network and infer the network structure and parameters by real gene expression data. By using the inferred network model, we may be able to discover the underlying gene regulatory mechanisms and therefore it helps to make useful predictions by computer simulation. The Boolean network model was first introduced by Kauffman [125, 126]. Advantages of this model can be found in Akutsu *et al.* [3], Kauffman [125, 126] and Shmulevich *et al.* [184, 185].

In this network model, each gene is regarded as a vertex of the network and is quantized into two levels only (express (0) or not-express (1)). Akutsu *et al.* [3] proposed the noisy Boolean networks together with an identification algorithm. In their model, they relax the requirement of consistency imposed by the Boolean functions. Regarding the effectiveness of a Boolean formalism, Shmulevich *et al.* [184, 185] proposed a PBN that can share the appealing rule-based properties of Boolean networks and it is robust in the presence of uncertainty. Their model is able to show a clear separation between different subtypes of gliomas as well as between different sarcomas by using multi-dimensional scaling. A logical representation of cell cycle regulation can also be found in Shmulevich *et al.* [184, 185]. However, it is widely recognized that reproducibility of measurements and between-slide variation are major issues. Moreover, genetic regulation also exhibits uncertainty on the biological level. Shmulevich also proposed a means of structural intervention method for controlling the stationary behavior in PBNs.

Boolean network modelling is commonly used for studying generic coarse-grained properties of large genetic networks without knowing specific quantitative details. Boolean network is deterministic, the only uncertainty is the initial starting state. Generally speaking, a Boolean network $G(V, \mathcal{F})$ consists of a set of nodes

$$V = \{v_1, v_2, \ldots, v_n\}$$

and $v_i(t)$ represents the state (0 or 1) of v_i at time t. A list of Boolean functions

$$\mathcal{F} = \{f^{(1)}, f^{(2)}, \ldots, f^{(n)}\}$$

represents the rules regulatory interaction between nodes:

$$v_i(t+1) = f^{(i)}(\mathbf{v}(t)), \quad i = 1, 2, \ldots, n,$$

where

$$\mathbf{v(t)} = (v_1(t), v_2(t), \ldots, v_n(t)).$$

In general, there may contain some unnecessary nodes in a Boolean function. For a Boolean function $f^{(j)}$, the variable $v_i(t)$ is said to be fictitious if

$$f^{(j)}(v_1(t), \ldots, v_{i-1}(t), 0, v_{i+1}(t), \ldots, v_n(t))$$
$$= f^{(j)}(v_1(t), \ldots, v_{i-1}(t), 1, v_{i+1}(t), \ldots, v_n(t))$$

for all possible values of

$$v_1(t), \ldots, v_{i-1}(t), v_{i+1}(t), \ldots, v_n(t).$$

We remark that when a Boolean network is used in the construction of underlying genetic networks, then n represents the number of genes under consideration, each vertex v_i represents the ith gene, and $v_i(t)$ represents the expression level of the ith gene at time t, taking either 0 or 1. The expression level of each gene is functionally related to that of other genes. Computational models that reveal these logical relations have been constructed in Bodnar [27], Mendoza et al. [154] and Huang et al. [116].

Standard Boolean networks are deterministic. However, in the biological aspect, an inherent determinism is not reasonable as it assumes an environment without uncertainty. The existence regularity of genetic function and interaction is caused by intrinsic self-organizing stability of the dynamical system instead of "hard-wired" logical rules, Shmulevich et al. [184]. In the empirical aspect, sample noise and relatively small amount of samples may cause incorrect results in logical rules. In order to overcome the deterministic rigidity of Boolean networks, the development of Probabilistic Boolean networks (PBNs) is essential. Not only PBN shares the appealing properties of Boolean networks, but also it is able to cope with uncertainty, including the data and model selection, Shmulevich et al. [184].

PBNs were firstly proposed by Shmulevich et al. [186] for genetic regulatory network. The model can be written as:

$$\mathcal{F}_i = \{f_j^{(i)}\}_{j=1,\ldots,l(i)},$$

where each predictor $f_j^{(i)}$ is a predictor determining the value of the gene v_i and $l(i)$ is the number of possible predictors for the gene v_i. It is clear that

$$\mathcal{F} = \bigcup_{i=1}^{n} \mathcal{F}_i.$$

We notice that when the number of possible PBN realization N is equal to 1 (i.e., $\prod_{i=1}^{n} l(i) = 1$), the PBN reduces to the standard Boolean network. Let $c_j^{(i)}$ be the probability that the j-th predictor, $f_j^{(i)}$, is chosen to predict the ith gene if $c_j^{(i)}$ is positive and this probability can be estimated by Coefficient of Determination (COD); Dougherty et al. (2000). Let us briefly describe COD here. Firstly, let $\epsilon_j^{(i)}$ be the optimal error achieved by $f_j^{(i)}$ and ϵ_i is the error of best estimate of ith gene in the absence of any conditional variable, then we have

$$\theta_j^{(i)} = \frac{\epsilon_i - \epsilon_j^{(i)}}{\epsilon_i}.$$

For all positive $\theta_j^{(i)}$, we can obtain $c_j^{(i)}$ by:

$$c_j^{(i)} = \frac{\theta_j^{(i)}}{\sum\limits_{k=1}^{l(i)} \{\theta_k^{(i)} : \theta_k^{(i)} > 0\}}.$$

Clearly, $c_j^{(i)}$ must satisfies

$$\sum_{j=1}^{l(i)} c_j^{(i)} = 1. \quad \text{for} \quad i = 1, \ldots, n.$$

For any given time point, the expression level of the ith gene is determined by one of the possible predictors $f_j^{(i)}$ for $1 \leq j \leq l(i)$. The probability of a transition from $\mathbf{v}(t)$ to $\mathbf{v}(t+1)$ can be obtained as

$$\prod_{i=1}^{n} \left[\sum_{k=1}^{l(i)} \left\{ c_k^{(i)} : f_k^{(i)}(\mathbf{v}(t)) = v_i(t+1) \right\} \right].$$

On the other hand, the level of influences from gene j to gene i can be estimated by

$$\begin{aligned} I_j(v_i) = \sum_{k=1}^{l(i)} \text{Prob}(f_k^{(i)}(v_1, \ldots, v_{j-1}, 0, v_{j+1}, \ldots, v_n) \\ \neq f_k^{(i)}(v_1, \ldots, v_{j-1}, 1, v_{j+1}, \ldots, v_n)) c_k^{(i)}. \end{aligned} \qquad (7.10)$$

Before evaluating either state transition probabilities or $I_j(v_i)$, we first need to obtain all the predictors $\bigcup_{i=1}^{n} \mathcal{F}_i$. We remark that for each set of $\mathcal{F}_i$ with $1 \leq i \leq n$, the maximum number of predictors is equal to 2^{2^n} as $1 \leq l(i) \leq 2^{2^n}$, it is also true for their corresponding probabilities

$$\{c_1^{(i)}, \ldots, c_{l(i)}^{(i)}\}.$$

It implies that the number of parameters in the PBN model is about $O(n2^{2^n})$. Obviously, the number of parameters increases exponentially with respect the number of genes n. Also, the COD used in obtaining $c_k^{(i)}$ must be estimated from the training data. Hence, it is almost impractical to apply this model due to either its model complexity or parameters imprecision owing to limited sample size. For the microarray-based analysis done by Kim $et~al.$ (2000), the number of genes in each set of $\mathcal{F}_i$ was kept to a maximum of three.

We note that PBN is a discrete-time process, the probability distribution of gene expression at time $t+1$ of the ith gene can be estimated by the gene expression of other n genes at time t via one-lag transition matrix. This is a Markov process framework. We consider the multivariate Markov chain model to infer the genetic network of n genes. In this network, no prior information on n genes relationships is assumed, our proposed model is used to uncover the underlying various gene relationships, including genes and genes cyclic or acyclic relationships. From our own model parameters, it is sufficient to uncover the gene regulatory network. However, one would like to have a fair performance comparison between PBNs and our model, we would like to illustrate using our model parameters to estimate some commonly used parameters in PBNs efficiently. In PBNs with n genes, there are n disjoint sets of predictors $\mathcal{F}_i$ and each of them is used for an unique gene sequence. In particular, for the d-th set of predictors $\mathcal{F}_d$, we notice that the possibility corresponding to each predictor $f_j^{(d)}$ can be obtained from our probability stationary vector and the detail is given as follows. We can estimate the conditional probability distribution $X_{i_1,\dots,i_n}^{(d)}$ for d output expression at base $t+1$ given by a set of genes input expression at base t, i.e.,

$$X_{i_1,\dots,i_n}^{(d)} = \mathrm{Prob}(V_{t+1}^{(d)} \mid V_t^{(k)} = E_{i_k} \text{ for } k = 1,\dots,n)$$

$$= \sum_{k=1}^{n} \lambda_{dk} P^{(dk)} E_{i_k} = \sum_{k=1}^{n} \lambda_{dk} P_{(\cdot,i_k)}^{(dk)}$$

where $i_k \in \{0,1\}$ and $P_{(\cdot,i)}^{(dk)}$ denote the i column of $P^{(dk)}$. Clearly, each probability vector $X_{i_1,\dots,i_n}^{(d)}$ is a unit vector and for each d, there are 2^n number of probability vectors we need to estimate. If $\lambda_{dj} = 0$ for some $j \in \{1,\dots,n\}$, it represents that the j-th gene does not have any influence to the d-th gene, and

$$X_{i_1,\dots,i_{j-1},0,i_{j+1},\dots,i_n}^{(d)} \equiv X_{i_1,\dots,i_{j-1},1,i_{j+1},\dots,i_n}^{(d)}$$

the number of estimated probability vectors could be reduced by half. After all the essential $X_{i_1,\dots,i_n}^{(d)}$ has been estimated, the probability $c_g^{(d)}$ of the predictor $f_g^{(d)}$ can be estimated by

$$c_g^{(d)} = \prod_{i_k \in \{0,1\}, k=1,\dots,n} X_{i_1,\dots,i_n}^{(d)}(f_g^{(d)}(i_1,\dots,i_n) + 1)$$

where

$$f_g^{(d)}(i_1,\dots,i_n) \in \{0,1\}$$

and $X_{i_1,\dots,i_n}(h)$ denotes the h entry of the vector $X_{i_1,\dots,i_n}$. If $c_g^{(d)} = 0$, the predictor $f_g^{(d)}$ does not exist and it should be eliminated. It is interesting to justify how the expression of ith gene is affected by the expression of jth gene, therefore, the degree of sensitivity from jth gene to ith gene can be estimated by equation (7.10) mentioned in previous section. We notice that there are two situations that $I_j(V_i) = 0$, Shmulevich et $al.$ [186], namely,

(i) If $\lambda_{ij} = 0$, then jth gene does not give any influence on ith gene.

(ii) The first two columns of the matrix $P^{(ij)}$ are identical, that means no matter the expression of jth gene is, the result of the probability vector is not affected.

7.6.1 An Example

Here we give an example to demonstrate the construction of our model parameters. We consider the following two binary sequences:

$$s_1 = \{0, 0, 1, 0, 0, 0, 0, 0, 1, 1, 0, 0\}$$

and

$$s_2 = \{1, 1, 0, 0, 1, 0, 0, 0, 0, 1, 0, 1\}.$$

We have the frequency matrices as follows:

$$F^{(11)} = \begin{pmatrix} 6 & 2 \\ 2 & 1 \end{pmatrix}, \qquad F^{(12)} = \begin{pmatrix} 5 & 3 \\ 2 & 1 \end{pmatrix},$$

$$F^{(21)} = \begin{pmatrix} 5 & 2 \\ 3 & 1 \end{pmatrix}, \qquad F^{(22)} = \begin{pmatrix} 4 & 3 \\ 3 & 1 \end{pmatrix}.$$

After normalization we have the transition probability matrices:

$$\hat{P}^{(11)} = \begin{pmatrix} \frac{3}{4} & \frac{2}{3} \\ \frac{1}{4} & \frac{1}{3} \end{pmatrix}, \qquad \hat{P}^{(12)} = \begin{pmatrix} \frac{5}{7} & \frac{3}{4} \\ \frac{2}{7} & \frac{1}{4} \end{pmatrix},$$

$$\hat{P}^{(21)} = \begin{pmatrix} \frac{5}{8} & \frac{2}{3} \\ \frac{3}{8} & \frac{1}{3} \end{pmatrix}, \qquad \hat{P}^{(22)} = \begin{pmatrix} \frac{4}{7} & \frac{3}{4} \\ \frac{3}{7} & \frac{1}{4} \end{pmatrix}.$$

Moreover we also have

$$\hat{V}_1 = (\frac{3}{4}, \frac{1}{4})^T$$

and

$$\hat{V}_2 = (\frac{7}{12}, \frac{5}{12})^T.$$

After solving the linear programming problem, the multivariate Markov model of the two binary sequences is given by

$$\begin{cases} V_{t+1}^{(1)} = 0.5\hat{P}^{(11)}V_t^{(1)} + 0.5\hat{P}^{(12)}V_t^{(2)} \\ V_{t+1}^{(2)} = 1.0\hat{P}^{(21)}V_t^{(1)} + 0.0\hat{P}^{(22)}V_t^{(2)}. \end{cases}$$

The conditional probability distribution vector $X_{0,0}^{(1)}$ can be estimated as:

$$X_{0,0}^{(1)} = 0.5\hat{P}^{(11)}(1,0)^T + 0.5\hat{P}^{(12)}(1,0)^T = (\frac{41}{56}, \frac{15}{56})^T.$$

We can obtain the rest of the vectors in the similar way and get:

$$X_{0,1}^{(1)} = (\frac{3}{4}, \frac{1}{4})^T, \quad X_{1,0}^{(1)} = (\frac{29}{42}, \frac{13}{42})^T$$

and

$$X_{1,1}^{(1)} = (\frac{17}{24}, \frac{7}{24})^T.$$

As $\lambda_{2,2} = 0$, therefore we have,

$$X_{0,0}^{(2)} = X_{0,1}^{(2)} = (\frac{5}{8}, \frac{3}{8})^T$$

and

$$X_{1,0}^{(2)} = X_{1,1}^{(2)} = (\frac{2}{3}, \frac{1}{3})^T.$$

From previous section, the probability $c_j^{(i)}$ can be obtained and the results are given in the Tables 7.3 and 7.4.

Table 7.3. The first sequence results.

v_1	v_2	$f_1^{(1)}$	$f_2^{(1)}$	$f_3^{(1)}$	$f_4^{(1)}$	$f_5^{(1)}$	$f_6^{(1)}$	$f_7^{(1)}$	$f_8^{(1)}$
0	0	0	0	0	0	0	0	0	0
0	1	0	0	0	0	1	1	1	1
1	0	0	0	1	1	0	0	1	1
1	1	0	1	0	1	0	1	0	1
$c_j^{(1)}$		0.27	0.11	0.12	0.05	0.08	0.04	0.04	0.02

v_1	v_2	$f_9^{(1)}$	$f_{10}^{(1)}$	$f_{11}^{(1)}$	$f_{12}^{(1)}$	$f_{13}^{(1)}$	$f_{14}^{(1)}$	$f_{15}^{(1)}$	$f_{16}^{(1)}$
0	0	1	1	1	1	1	1	1	1
0	1	0	0	0	0	1	1	1	1
1	0	0	0	1	1	0	0	1	1
1	1	0	1	0	1	0	1	0	1
$c_j^{(1)}$		0.1	0.04	0.04	0.02	0.03	0.01	0.02	0.01

For instance,

$$c_6^{(1)} = [X_{0,0}^{(1)}]_1 \times [X_{0,1}^{(1)}]_2 \times [X_{1,0}^{(1)}]_1 \times [X_{1,1}^{(1)}]_2$$

$$= \frac{41}{56} \times \frac{1}{4} \times \frac{29}{42} \times \frac{7}{24} = 0.04.$$

Because of $\lambda_{22} = 0$, the set of predictors for the second sequence can reduce significantly.

From Tables 7.3 and 7.4, the level of sensitivity $I_j(v_i)$ can be obtained by direct calculation. For example,

Table 7.4. The second sequence results.

v_1	v_2	$f_1^{(2)}$	$f_2^{(2)}$	$f_3^{(2)}$	$f_4^{(2)}$
0	—	0	0	1	1
1	—	0	1	0	1
$c_j^{(2)}$		0.42	0.2	0.25	0.13

$$
\begin{aligned}
I_1(v_1) &= 0(0.27) + \tfrac{1}{2}(0.11) + \tfrac{1}{2}(0.12) + 0.05 \\
&\quad + \tfrac{1}{2}(0.08) + 0(0.04) + 0.04 + \tfrac{1}{2}(0.02) \\
&\quad + \tfrac{1}{2}(0.1) + 0.04 + 0(0.04) + \tfrac{1}{2}(0.02) \\
&\quad + (0.03) + \tfrac{1}{2}(0.01) + \tfrac{1}{2}(0.02) + 0(0.01) \\
&= 0.4.
\end{aligned}
$$

and we have

$$
I_2(v_1) = 0.4, \quad I_1(v_2) = 0.45 \quad \text{and} \quad I_2(v_2) = 0.
$$

According to the calculated values $I_i(v_j)$, we know that the first sequence somehow determine the second sequence. However, this phenomena is already illustrated by the fact that $\lambda_{22} = 0$ ($\lambda_{21} = 1$) in the multivariate Markov chain model.

7.6.2 Fitness of the Model

The multivariate Markov chain model presented here is a stochastic model. Given all the state vectors $\mathbf{V}_t^{(k)}$ with $k = 1, \ldots, n$, the state probability distribution $\mathbf{V}_{t+1}^{(k)}$ can be estimated by using (7.1). According to this state probability distribution, one of the prediction methods for the jth sequence at time $t + 1$ can be taken as the state with the maximum probability, i.e.,

$$
\hat{\mathbf{V}}(t + 1) = j, \quad \text{if } [\hat{\mathbf{V}}(t + 1)]_i \leq [\hat{\mathbf{V}}(t + 1)]_j \text{ for all } 1 \leq i \leq 2.
$$

By making use of this treatment, our multivariate Markov chain model can be used to uncover the rules (build a truth table) for PBNs. With higher prediction accuracy, we have more confidence that the true genetic networks are uncovered by our model. To evaluate the performance and effectiveness, the prediction accuracy of all individual sequences r and the joint sequences R are defined respectively as follow:

$$
r = \frac{1}{nT} \times \sum_{i=1}^{n} \sum_{t=1}^{T} \delta_t^{(i)} \times 100\%,
$$

where

$$
\delta_t^{(i)} = \begin{cases} 1, & \text{if } \hat{\mathbf{v}}_i(t) = \mathbf{v}_i(t) \\ 0, & \text{otherwise.} \end{cases}
$$

and

$$R = \frac{1}{T} \times \sum_{t=1}^{T} \delta_t \times 100\%,$$

where

$$\delta_t = \begin{cases} 1, & \text{if } \hat{\mathbf{v}}_i(t) = \mathbf{v}_i(t) \quad \text{for all } 1 \le i \le n \\ 0, & \text{otherwise.} \end{cases}$$

Here T is the length of the data sequence. From the values of r and R, the accuracy of network realization for an individual sequence and for a whole set of sequences could be determined respectively. In this subsection, we test our multivariate Markov chain model for yeast data sequence.

Test with the Gene Expression Data of Yeast

Genome transcriptional analysis has been shown to be important in medicine, and etiology as well as in bioinformatics. One of the applications of genome transcriptional analysis is the eukaryotic cell cycle in yeast. The fundamental periodicity in eukaryotic cell cycle includes the events of DNA replication, chromosome segregation and mitosis. Hartwell and Kastan [105] suggested that improper cell cycle regulation may lead to genomic instability, especially in etiology of both hereditary and spontaneous cancers, Wang et al. [205]; Hall and Peters [104]. Eventually, it is believed to play one of the important roles in the etiology of both hereditary and spontaneous cancers. Genome transcriptional analysis helps in exploring the cell cycle regulation and the mechanism behind the cell cycle. Raymond et al. [176] examined the present of cell cycle-dependent periodicity in 6220 transcripts and found that cell cycles appear in about 7% of transcripts. Those transcripts are then extracted for further examination. When the time course was divided into early G1, late G1, S, G2 and M phase, the result showed that more than 24% of transcripts are directly adjacent to other transcripts in the same cell cycle phase. The division is based on the size of the bugs and the cellular position of the nucleus. Further investigating result on those transcripts also indicated that more than half are affected by more than one cell cycle-dependent regulatory sequence.

In our study, we use the data set selected from Yeung and Ruzzo [213]. In the discretization, if an expression level is above (below) its standard deviation from the average expression of the gene, it is over-expressed (under-expressed) and the corresponding state is 1 (0). Our main goal is to find out the relationship in 213 well-known yeast transcripts with cell cycle in order to illustrate the ability of our proposed model. This problem can be solved by using a PBN theoretically. However, there are problems in using PBNs in practice. It is clearly that the method of COD is commonly used to estimate the probabilities of each predictor $c_g^{(d)}$ for transcript d. Unfortunately, owing to limited time points of the expression level of each gene (there are only 17 time points for the yeast data set), it is almost impossible to find a value of $c_g^{(d)}$ which is strictly greater than that of the best estimation in the absence of any conditional variables. Therefore, most of the transcripts do not have any predictor and it leads to all of the parameters in PBN are impossible to be estimated. Moreover, PBN seems to be unable to model a set of genes when n is quite large. Nir *et al.* [162] suggested Bayesian networks can infer a genetic network successfully, but it is unable to infer a genetic network with cell cycle relationship. Ott *et al.* [165] also suggested that even if in a acyclic genetic network with constraints situation, the number of genes in Bayesian networks should not be greater than 40 if BNRC score are used. Kim *et al.* [129] proposed a dynamic Bayesian network which can construct of cyclic regulations for medium time-series, but still it cannot handle a large network. Here, we use the multivariate Markov chain model for training the yeast data. The construction of a multivariate Markov chain model for such data set only requires around 0.1 second. We assume that there is no any prior knowledge about the genes. In the construction of the multivariate Markov chain model, each target gene can be related to other genes. Based on the values of λ_{ij} in our model, one can determine the occurrence of cell cycle in jth transcript, i.e., in a set of transcripts, there present a inter-relationship of any jth transcript in this set. Based on the built multivariate Markov chain model, 93% of transcripts possibly involves in some cell cycles were found. Some of the results are shown in Table 7.5.

Table 7.5. Results of our multivariate Markov chain model.

No.	Name of target transcript	Cell cycle phase	Length of cell cycle	Related transcripts (its phase λ_{ij}, level of influence)
(1)	YDL101c	late G1	1	YMR031c(1,1.00,1.00)
(2)	YKL113c	late G1	2	YDL018c (2,0.50,0.50)
				YOR315w(5,0.50,0.50)
				YML027w(2,0.33,0.39)
				YJL079c(5,0.33,0.38)
(3)	YLR121c	late G1	3	YPL158c(1,0.33,0.42)
				YDL101c(2,0.33,0.43)
				YKL069w(4,0.33,0.43)
				YER001w(3,0.50,0.50)
(4)	YLR015w	early G1	4	YKL113c(2,1.00,0.88)

In Table 7.5, the first column indicates the number of data set we display. The second column gives the name of target transcript. The third column shows which phase the target gene belongs to. The fourth column shows the most possibly cell cycle length of the target transcript. Finally, the last column displays the name of required transcripts for predicting the target transcript, the corresponding phase of required transcripts, their corresponding weights λ_{ij} in the model, as well as an estimated value of the level of influence from related transcript to the target transcript. Although the level of influence can be estimated based on our model parameters, its computational cost in the PBN method increases exponentially respect to the value of n.

We find in Table 7.5 that the weighting λ_{ij} provides a reasonable measure for the level of influence. Therefore the proposed method can estimate the level of influence very efficiently. Finally, we present in Table 7.6 the prediction results of different lengths of cell cycles for the whole data set and the results show that the performance of the model is good.

Table 7.6. Prediction results.

Length of cell cycle phases required	No. of occurrence in this type of cell cycle	Average prediction accuracy	Example in Table 7.5
1	5 %	86 %	(1)
2	9 %	87 %	(2)
3	9 %	83 %	(3)
4	70 %	86 %	(4)

Further research can be done in gene perturbation and intervention. We note that a PBN allows uncertainty of inter-gene relations in the dynamic process and it will evolve only according to certain fixed transition probabilities. However, there is no mechanism to control this process so as to achieve some desirable states. To facilitate PBNs to evolve towards some desirable directions, intervention has been studied. It has been shown that given a target state, one can facilitate the transition to it by toggling the state of a particular gene from on to off or vice-versa Shmulevich et al. [187]. But making a perturbation or a forced intervention can only be applied at one time point. The dynamics of the system thereafter still depends on the network itself. Thus the network may eventually return to some undesirable state after a number of steps. Another way to tackle this problem is to by use structural intervention to change the stationary behavior of the PBNs Shmulevich et al. [185]. This approach constitutes transient intervention. It involves the structural intervention and therefore it will be more permanent. By using the proposed multivariate Markov chain model, it is possible to formulate the gene intervention problem as a linear control model. To increase the likelihood of transitions to a desirable state, more auxiliary variables can be introduced in the system Datta et al. [81]. Moreover, costs can be assigned to the control inputs and also the states researched such that higher terminal costs are assigned to those undesirable states. The objective here is to achieve a target state probability distribution with a minimal control cost. The model can be formulated as a minimization problem with integer variables and continuous variables, Zhang et al [218].

7.7 Extension to Higher-order Multivariate Markov Chain

In this section, we present our higher-order multivariate Markov chain model for modelling multiple categorical sequences based on the models in Sections 6.2 and 7.2. We assume that there are s categorical sequences with order n and each has m possible states in $\mathcal{M}$. In the extended model, we assume that the state probability distribution of the jth sequence at time $t = r+1$ depends on the state probability distribution of all the sequences (including itself) at times $t = r, r-1, \ldots, r-n+1$. Using the same notations as in the previous two subsections, our proposed higher-order (nth-order) multivariate Markov chain model takes the following form:

$$\mathbf{x}_{r+1}^{(j)} = \sum_{k=1}^{s} \sum_{h=1}^{n} \lambda_{jk}^{(h)} P_h^{(jk)} \mathbf{x}_{r-h+1}^{(k)}, \quad j = 1, 2, \ldots, s \tag{7.11}$$

where

$$\lambda_{jk}^{(h)} \geq 0, \quad 1 \leq j, k \leq s, \quad 1 \leq h \leq n \tag{7.12}$$

and

$$\sum_{k=1}^{s} \sum_{h=1}^{n} \lambda_{jk}^{(h)} = 1, \quad j = 1, 2, \ldots, s.$$

The probability distribution of the jth sequence at time $t = r + 1$ depends on the weighted average of $P_h^{(jk)} \mathbf{x}_{r-h+1}^{(k)}$. Here $P_h^{(jk)}$ is the hth-step transition probability matrix which describes the hth-step transition from the states in the kth sequence at time $t = r - h + 1$ to the states in the jth sequence at time $t = r + 1$ and $\lambda_{jk}^{(h)}$ is the weighting of this term.

From (7.11), if we let

$$\mathbf{X}_r^{(j)} = (\mathbf{x}_r^{(j)}, \mathbf{x}_{r-1}^{(j)}, \ldots, \mathbf{x}_{r-n+1}^{(j)})^T \quad \text{for} \quad j = 1, 2, \ldots, s$$

be the $nm \times 1$ vectors then one can write down the following relation in matrix form:

$$\mathbf{X}_{r+1} \equiv \begin{pmatrix} \mathbf{X}_{r+1}^{(1)} \\ \mathbf{X}_{r+1}^{(2)} \\ \vdots \\ \mathbf{X}_{r+1}^{(s)} \end{pmatrix} = \begin{pmatrix} B^{(11)} & B^{(12)} & \cdots & B^{(1s)} \\ B^{(21)} & B^{(22)} & \cdots & B^{(2s)} \\ \vdots & \vdots & \vdots & \vdots \\ B^{(s1)} & B^{(s2)} & \cdots & B^{(ss)} \end{pmatrix} \begin{pmatrix} \mathbf{X}_r^{(1)} \\ \mathbf{X}_r^{(2)} \\ \vdots \\ \mathbf{X}_r^{(s)} \end{pmatrix} \equiv Q\mathbf{X}_r$$

where

$$B^{(ii)} = \begin{pmatrix} \lambda_{ii}^{(n)} P_n^{(ii)} & \lambda_{ii}^{(n-1)} P_{n-1}^{(ii)} & \cdots & \lambda_{ii}^{(2)} P_2^{(ii)} & \lambda_{ii}^{(1)} P_1^{(ii)} \\ I & 0 & \cdots & 0 & 0 \\ 0 & I & \cdots & 0 & 0 \\ \vdots & \ddots & \ddots & \ddots & 0 \\ 0 & \cdots & 0 & I & 0 \end{pmatrix}_{mn \times mn}$$

and if $i \neq j$ then

$$B^{(ij)} = \begin{pmatrix} \lambda_{ij}^{(n)} P_n^{(ij)} & \lambda_{ij}^{(n-1)} P_{n-1}^{(ij)} & \cdots & \lambda_{ij}^{(2)} P_2^{(ij)} & \lambda_{ij}^{(1)} P_1^{(ij)} \\ 0 & 0 & \cdots & 0 & 0 \\ 0 & 0 & \cdots & 0 & 0 \\ \vdots & \ddots & \ddots & \ddots & 0 \\ 0 & \cdots & 0 & 0 & 0 \end{pmatrix}_{mn \times mn} .$$

We note that each column sum of Q is not necessary equal to one but each column sum of $P_h^{(jk)}$ is equal to one. We have the following propositions.

Proposition 7.3. *If $\lambda_{jk}^{(h)} > 0$ for $1 \leq j, k \leq s$ and $1 \leq h \leq n$, then the matrix Q has an eigenvalue equal to one and the eigenvalues of Q have modulus less than or equal to one.*

Proposition 7.4. *Suppose that $P_h^{(jk)}$ $(1 \leq j, k \leq s, 1 \leq h \leq n)$ are irreducible and $\lambda_{jk}^{(h)} > 0$ for $1 \leq j, k \leq s$ and $1 \leq h \leq n$. Then there is a vector*

$$\mathbf{X} = (\mathbf{X}^{(1)}, \mathbf{X}^{(2)}, \dots, \mathbf{X}^{(s)})^T$$

with

$$\mathbf{X}^{(j)} = (\mathbf{x}^{(j)}, \mathbf{x}^{(j)}, \dots, \mathbf{x}^{(j)})^T$$

such that

$$\mathbf{X} = Q\mathbf{X} \quad \text{and} \quad \mathbf{1x}^{(j)} = 1, \quad \text{for} \quad 1 \leq j \leq s$$

$\mathbf{1} = (1, 1, \dots, 1)$ *of length* m.

The transition probabilities P_{jk}^h can be estimated by counting the transition frequency as described in Section 6.2 of Chapter 6 and Section 7.2. Moreover, we note that $\mathbf{X}$ is not a probability distribution vector, but $\mathbf{x}^{(j)}$ is a probability distribution vector. The above proposition suggests one possible way to estimate the model parameters $\lambda_{ij}^{(h)}$. The key idea is to find $\lambda_{ij}^{(h)}$ which minimizes $||Q\hat{\mathbf{x}} - \hat{\mathbf{x}}||$ under certain vector norm $|| \cdot ||$. The estimation method is similar to those in Chapter 6. The proofs of Propositions 7.3 and 7.4 and detailed examples of demonstration with an application in production planning can be found in Ching et al. [65].

7.8 Summary

In this chapter, we present the a multivariate Markov chain model with estimation methods for the model parameters based on solving linear programming problem. The model has been applied to multi-product demand estimation problem, credit rating problem, multiple DNA sequences and genetic networks. We also extend the model to a higher-order multivariate Markov chain model. Further research can be done on the following issues.

(i) New estimation methods when there are missing data in the given sequences.

(ii) The case when the model parameters λ_{ij} are allowed to take negative values. The treatment can be similar to the discussion in Section 6.4.

8

Hidden Markov Chains

8.1 Introduction

Hidden Markov models (HMMs) have been applied to many real-world applications. Very often HMMs only deal with the first-order transition probability distribution among the hidden states, see for instance Section 1.4. Moreover, the observable states are affected by the hidden states but not vice versa. In this chapter, we study both higher-order hidden Markov models and interactive HMM in which the hidden states are directly affected by the observed states. We will also develop estimation methods for the model parameters in both cases.

The remainder of this chapter is organized as follows. In Section 8.2, we present a higher-order hidden Markov model. In Section 8.3, we discuss an interactive HMM. In Section 8.4, we discuss a double higher-order hidden Markov models. Finally, a summary will be given to conclude this chapter in Section 8.5.

8.2 Higher-order HMMs

In this section, we present a higher-order Hidden Markov Model (HMM) and the model is applied to modeling DNA sequences, see Ching et al. [61]. HMMs have become increasingly popular in the last few decades. Since HMMs are very rich in mathematical structure, they can form the theoretical basis in a wide range of applications such as the DNA sequences [135], speech recognition [173] and computer version [39]. A standard HMM is usually characterized by the following elements [173]:

(i) N, the number of states in the model. Although the states are hidden, for many practical applications, very often, there is physical significance to the states. We denote the individual states as

$$\mathbf{S} = \{S_1, S_2, \ldots, S_N\},$$

and the state at the length t as q_t.

(ii) M, the number of distinct observation symbols (or state) for the hidden states. The observation symbols correspond to the physical output of the system being modeled. We denote the individual symbols as

$$\mathbf{V} = \{v_1, v_2, \ldots, v_M\}.$$

(iii) The state transition probability distribution

$$A = \{a_{ij}\}$$

where

$$a_{ij} = P(q_{t+1} = S_j | q_t = S_i), \quad 1 \le i, j \le N.$$

(iv) The observation probability distribution in state j, $B = \{b_j(k)\}$, where

$$b_j(k) = P(O_t = v_k | q_t = S_j), \quad 1 \le j \le N, 1 \le k \le M.$$

(v) The initial state distribution $\Pi = \{\pi_i\}$ where

$$\pi_i = P(q_1 = S_i), \quad 1 \le i \le N.$$

Given appropriate values of N, M, A, B and Π, the HMM can be used as a generator to give an observation sequence

$$O = O_1 O_2 \ldots O_T$$

where each observation O_t is one of the symbols from $\mathbf{V}$, and T is the number of observations in the sequence. For simplicity, we use the compact notation

$$\Lambda = (A, B, \Pi)$$

to indicate the complete parameter set of the HMM. According to the above specification, very often a first order Markov process is used in modeling the transitions among the hidden states in a HMM. In the DNA sequence analysis, higher-order Markov models have been used to model the transitions among the observable states, see [28, 100]. An mth order Markov process is a stochastic process where each event depends on the previous m events. It is believed that higher-order Markov model (in the hidden layer) can better capture a number of data sequences such as the DNA sequences. The main aim of this paper is to develop higher-order HMMs (higher-order Markov model for the hidden states). The main difference between the traditional HMM and a higher-order HMM is that in the hidden layer, the state transition probability is governed by the mth order higher-order Markov model

$$a_{i_{t-m+1}, \ldots, i_{t+1}} = P(q_{t+1} = S_{i_{t+1}} | q_t = S_{i_t}, \ldots, q_{t-m+1} = S_{i_{t-m+1}}).$$

We assume that the distribution Π of initial m states is given by

$$\pi_{i_1,i_2,\ldots,i_m} = P(q_1 = S_{i_1}, q_2 = S_{i_2}, \ldots, q_m = S_{i_m}).$$

Here we will present solution to the three problems for higher-order HMMs. Recall that they are practical problems in the traditional HMMs (see Section 1.4).

- **Problem 1** Given the observation sequence

$$O = O_1 O_2 \ldots O_T$$

  and a higher-order HMM, how to efficiently compute the probability of the observation sequence?
- **Problem 2** Given the observation sequence

$$O = O_1 O_2 \ldots O_T$$

  and a higher-order HMM, how to choose a corresponding state sequence

$$Q = q_1 q_2 \ldots q_T$$

  which is optimal in certain sense (e.g. in the sense of maximum likelihood)?
- **Problem 3** Given the observation sequence

$$O = O_1 O_2 \ldots O_T$$

  and a higher-order HMM, how to choose the model parameters?

8.2.1 Problem 1

For Problem 1, we calculate the probability of the observation sequence,

$$O = O_1 O_2 \ldots O_T,$$

given the higher-order HMM, i.e., $P[O|\Lambda]$. One possible way of doing this is through enumerating each possible state sequence of length T. However, this calculation is computationally infeasible even for small values of T and N. We apply the forward-backward procedure [14] to calculate this probability of the observation sequence. We define the forward variable

$$\alpha_t(i_{t-m+1}, \ldots, i_t)$$

as follows:

$$\alpha_t(i_{t-m+1}, \ldots, i_t) = P(O_1, \ldots, O_t, q_{t-m+1} = S_{i_{t-m+1}}, \ldots, q_t = S_{i_t}|\Lambda),$$

where $m \le t \le T$, i.e., the conditional probability that the subsequence of the first t observations and the subsequence of last m hidden states ending at time t are equal to

$$v_1 \ldots v_t \quad \text{and} \quad S_{i_{t-m+1}} \ldots S_{i_t}$$

respectively, are given by the model parameters Λ. We see that if we can obtain the values of

$$\alpha_T(i_{T-m+1}, \ldots, i_T) \quad \forall \quad i_{T-m+1}, \ldots, i_T,$$

then it is obvious that $P[O|\Lambda]$ can be obtained by summing up all the values of

$$\alpha_T(i_{T-m+1}, \ldots, i_T).$$

It is interesting to note that the values of $\alpha_T(i_{T-m+1}, \ldots, i_T)$ can be obtained by the following recursive equation and the details are given as follows:

(F1) Initialization: $\alpha_m(i_1, i_2, \ldots, i_m) = \pi_{i_1, i_2, \ldots, i_m} \cdot \prod_{j=1}^{m} b_{i_j}(v_j).$

(F2) Recursive Equation: $\alpha_{t+1}(i_{t-m+2}, i_{t-m+3}, \ldots, i_{t+1}) =$

$$\sum_{i_{t-m+1}=1}^{N} \alpha_t(i_{t-m+1}, \ldots, i_t) \cdot P(O_{t+1}|\Lambda, q_{t+1} = S_{i_{t+1}}) \cdot$$
$$P(q_{t+1} = S_{i_{t+1}}|\Lambda, q_{t-m+1} = S_{i_{t-m+1}}, \ldots, q_t = S_{i_t}))$$
$$= \sum_{i_{t-m+1}=1}^{N} \alpha_t(i_{t-m+1}, \ldots, i_t) \cdot a_{i_{t-m+1}i_t, i_{t+1}} b_{i_{t+1}}(v_{t+1}).$$

(F3) Termination: $P(O|\Lambda) = \sum_{i_{T-m+1}, \ldots, i_T=1}^{N} \alpha_T(i_{T-m+1}, \ldots, i_T).$

The initiation step calculates the forward probabilities as the joint probability of hidden states and initial observations. The recursion step, which is the main part of the forward calculation. Finally, the last step gives the desired calculation of $P[O|\Lambda]$ as the sum of the terminal forward variables $\alpha_T(i_{T-m+1}, \ldots, i_T)$. In a similar manner, a backward variable $\beta_t(i_1, i_2, \ldots, i_m)$ can be defined as follows: $\beta_t(i_1, i_2, \ldots, i_m) =$

$$P(O_{t+m} \ldots O_T|q_t = S_{i_t}, \ldots, q_{t+m-1} = S_{i_{t+m-1}}, \Lambda), \ 0 \le t \le T - m.$$

(B1) Initialization: $\beta_{T-t}(i_1, \ldots, i_m) = 1, \ 0 \le t \le m - 1, \ 1 \le i_1, \ldots, i_m \le N.$

(B2) Recursive equation: $\beta_t(i_1, i_2, \ldots, i_m) =$

$$\sum_{i_{t+m}=1}^{N} P(O_{t+m+1} \ldots O_T|q_{t+1} = S_{i_{t+1}}, \ldots, q_{t+m-1} = S_{i_{t+m-1}}, q_{t+m} = S_{i_{t+m}}, \Lambda) \cdot$$
$$P(O_{t+m}|q_{t+m} = S_{i_{t+m}}, \Lambda) \cdot P(q_{t+m} = S_{i_{t+m}}|q_t = S_{i_t}, \ldots, q_{t+m-1} = S_{i_{t+m-1}}, \Lambda)$$
$$= \sum_{k=1}^{N} b_k(O_{t+m})\beta_{t+1}(i_2, \ldots, i_m, k) \cdot a_{i_2, \ldots, i_m, k}.$$

The initialization step arbitrarily defines $\beta_{T-t}(i_1, i_2, \ldots, i_m)$ to be 1. The induction step of the backward calculation is similar to the forward calculation.

8.2.2 Problem 2

In Problem 2, we attempt to uncover the whole hidden sequence give the observations, i.e. to find the most likely state sequence. In practical situations, we use an optimality criteria to solve this problem as good as possible. The most widely used criterion is to find the best sequence by maximizing $P[Q|\Lambda, O]$. This is equivalent to maximize $P(Q, O|\Lambda)$. We note that

$$P(Q|\Lambda, O) = \frac{P(Q, O|\Lambda)}{P(O|\Lambda)}.$$

Viterbi algorithm [204] is a technique for finding this "best" hidden sequence $Q = \{q_1, q_2, \ldots, q_T\}$ for a given observation sequence $O = \{O_1, O_2, \ldots, O_T\}$. Here we need to define the following quantity:

$$\delta_t(i_{t-m+1}, \ldots, i_t) = \max_{q_1, \ldots, q_{t-m}} P(q_1 = S_{i_1}, \ldots, q_t = S_{i_t}, O_1, \ldots, O_t|\Lambda),$$

for $m \le t \le T$ and $\delta_t(i_{t-m+1}, \ldots, i_t)$ is the best score (highest probability) along a single best state sequence at time t, which accounts for the first t observations and ends in state S_{i_t}. By induction, we have

$$\delta_{t+1}(i_{t-m+2}, \ldots, i_{t+1})$$
$$= \max_{1 \le q_{t-m+1} \le N} \{\delta_t(i_{t-m+1}, \ldots, i_t) \cdot a_{i_{t-m+1}, \ldots, i_{t+1}}\} \cdot b_{i_{t+1}}(O_{t+1}). \quad (8.1)$$

To retrieve the state sequence, ones needs to keep track of the argument which maximized (8.1) for each t and $i_{t-m+1}, \ldots, i_t$. this can be done via the array $\Delta_{t+1}(i_{t-m+2}, \ldots, i_{t+1})$. The complete procedure for finding the best state sequence is as follows:

(U1) Initialization:

$$\delta_m(i_1, \ldots, i_m) = P(q_1 = S_{i_1}, \ldots, q_m = S_{i_m}, O_1, \ldots, O_m|\Lambda)$$
$$= P(q_1 = S_{i_1}, \ldots, q_m = S_{i_m}|\Lambda) \cdot \prod_{j=1}^{m} P(O_j|\Lambda, q_j = S_{i_j})$$
$$= \pi_{i_1, i_2, \ldots, i_m} \prod_{j=1}^{m} b_{i_j}(v_j), \quad 1 \le i_1, i_2, \ldots, i_m \le N.$$

We also set $\Delta_m(i_1, \ldots, i_m) = 0$.

(U2) Recursion:

$$\delta_{t+1}(i_{t-m+2}, \ldots, i_{t+1})$$
$$= \max_{q_1, \ldots, q_{t-m+1}} P(q_{t+1} = S_{i_{t+1}}, O_{t+1}|\Lambda, q_1 = i_1, \ldots, q_t = i_t, O_1, \ldots, O_t) \cdot$$
$$P(q_1 = S_{i_1}, \ldots, q_t = S_{i_t}, O_1, \ldots, O_t|\Lambda)$$

$$= \max_{1 \leq q_{t-m+1} \leq N} \delta_t(i_{t-m+1}, \ldots, i_t) \cdot$$
$$P(O_{t+1}|\Lambda, q_1 = S_{i_1}, \ldots, q_{t+1} = S_{i_{t+1}}, O_1, \ldots, O_t) \cdot$$
$$P(q_{t+1} = S_{i_{t+1}}|\Lambda, q_1 = S_{i_1}, \ldots, q_t = S_{i_t}, O_1, \ldots, O_t)$$
$$= \max_{1 \leq q_{t-m+1} \leq N} \delta_t(i_{t-m+1}, \ldots, i_t) \cdot P(O_{t+1}|\Lambda, q_{t+1} = S_{i_{t+1}}) \cdot$$
$$P(q_{t+1} = S_{i_{t+1}}|\Lambda, q_{t-m+1} = S_{i_{t-m+1}}, \ldots, q_t = S_{i_t})$$
$$= \max_{1 \leq q_{t-m+1} \leq N} \{\delta_t(i_{t-m+1}, \ldots, i_t) \cdot a_{i_{t-m+1}, \ldots, i_{t+1}}\} \cdot b_{i_{t+1}}(v_{t+1}).$$

For $m + 1 \leq t \leq T$ and $1 \leq i_{t+1} \leq N$, we have

$$\Delta_{t+1}(i_{t-m+2}, \ldots, i_{t+1})$$
$$= \mathrm{argmax}_{1 \leq q_{t-m+1} \leq N}\{\delta_t(i_{t-m+1}, \ldots, i_t) \cdot a_{i_{t-m+1}, \ldots, i_{t+1}}\}.$$

(U3) Termination

$$P^* = \max_{1 \leq q_{T-m+1}, \ldots, q_T \leq N}\{\delta_{q_{T-m+1}, \ldots, q_T}\}$$

$$(q_{T-m+1}^*, \ldots, q_T^*) = \mathrm{argmax}_{1 \leq q_{T-m+1}, \ldots, q_T \leq N}\{\delta_{q_{T-m+1}, \ldots, i_T}\}$$

8.2.3 Problem 3

In Problem 3, we attempt to adjust the model parameters Λ by maximizing the probability of the observation sequence given the model. Here we choose Λ such that $P[O|\Lambda]$ is maximized with the assumption that the distribution Π of the initial m states is known by using the EM algorithm. Define

$$C(\Lambda, \overline{\Lambda}) = \sum_Q P(Q|O, \Lambda) \log P(O, Q|\overline{\Lambda}).$$

The EM algorithm includes two main steps, namely E-step, calculating the function $C(\Lambda, \overline{\Lambda})$ and the M-step, maximizing $C(\Lambda, \overline{\Lambda})$ with respect to $\overline{\Lambda}$. Now, we define $\epsilon_t(i_1, i_2, \ldots, i_{m+1})$ as follows:

$$\epsilon_t(i_1, i_2, \ldots, i_{m+1}) = P(q_t = S_{i_1}, q_{t+1} = S_{i_2}, \ldots, q_{t+m} = S_{i_{m+1}}|O, \Lambda).$$

We can write down the expression of $\epsilon_t(i_1, i_2, \ldots, i_{m+1})$ in terms of $\alpha(\cdot)$ and $\beta(\cdot)$ that are computed in the previous two sub-sections:

$$\epsilon_t(i_1, i_2, \ldots, i_{m+1})$$
$$= b_{i_{m+1}}(O_{t+m})P[O_{t+m+1} \ldots O_T|q_{t+1} = S_{i_2}, \ldots, q_{t+m} = S_{i_{m+1}}, \Lambda] \cdot$$
$$P(q_{t+m} = S_{i_{m+1}}|q_t = S_{i_1}, q_{t+1} = S_{i_2}, \ldots, q_{t+m-1} = S_{i_m}, \Lambda) \cdot$$
$$P[O_1 O_2 \ldots O_{t+m-1}, q_t = S_{i_1}, q_{t+1} = S_{i_2}, \ldots, q_{t+m-1} = S_{i_m}|\Lambda)$$
$$= \alpha_{t+m-1}(i_1, i_2, \ldots, i_m)a_{i_1, \ldots, i_{m+1}}b_{i_{m+1}}(O_{t+m})\beta_{t+1}(i_2, i_3, \ldots, i_{m+1}).$$

Therefore we obtain

$$\epsilon_t(i_1, i_2, \ldots, i_{m+1}) = P(q_t = S_{i_1}, q_{t+1} = S_{i_2}, \ldots, q_{t+m} = S_{i_{m+1}} | O, \Lambda)$$

$$= \frac{\alpha_{t+m-1}(i_1, i_2, \ldots, i_m) a_{i_1, \ldots, i_{m+1}} b_{i_{m+1}}(O_{t+m}) \beta_{t+1}(i_2, i_e, \ldots, i_{m+1})}{P[O|\Lambda]}.$$

Next we define

$$\gamma_t(i_1, i_2, \ldots, i_k) = \sum_{i_{k+1}=1}^{N} \cdots \sum_{i_{m+1}=1}^{N} \epsilon_t(i_1, i_2, \ldots, i_{m+1}).$$

If we sum $\epsilon_t(i_1, i_2, \ldots, i_{m+1})$ over the index t, we get a quantity which can be interpreted as the expected number of times that state sequence $S_{i_1} S_{i_2} \cdots S_{i_{m+1}}$ occurred. Similarly, if we sum $\gamma_t(i_1, i_2, \ldots, i_m)$ over t, we get a quantity which can be interpreted as the expected number of times that state sequence $S_{i_1} S_{i_2} \cdots S_{i_m}$ occurred. Hence, a set of re-estimation formulae is given as follows:

$$
\left\{
\begin{aligned}
\gamma_t(i_1) &= \sum_{i_2=1}^{N} \sum_{i_3=1}^{N} \cdots \sum_{i_{m+1}=1}^{N} \epsilon_t(i_1, i_2, \ldots, i_{m+1}), \\
\gamma_t(i_1, i_2) &= \sum_{i_3=1}^{N} \cdots \sum_{i_{m+1}=1}^{N} \epsilon_t(i_1, i_2, \ldots, i_{m+1}), \\
&\vdots \\
\gamma_t(i_1, i_2, \ldots, i_m) &= \sum_{i_{m+1}=1}^{N} \epsilon_t(i_1, i_2, \ldots, i_{m+1}), \\
\overline{\pi_{i_1}} &= \gamma_1(i_1), \\
\overline{\pi_{i_1 i_2}} &= \gamma_1(i_1, i_2), \\
&\vdots \\
\overline{\pi_{i_1 i_2 \ldots i_m}} &= \gamma_1(i_1, i_2, \ldots, i_m), \\
A_{i_1 i_2 \ldots i_{m+1}} &= \sum_{t=1}^{T-m} \epsilon_t(i_1, i_2, \ldots, i_{m+1}), \\
A_{i_1 i_2 \ldots i_m} &= \sum_{i_{m+1}=1}^{N} A_{i_1 i_2 \ldots i_{m+1}}, \\
\overline{a_{i_1, \ldots, i_{m+1}}} &= A_{i_1 i_2 \ldots i_{m+1}} / \sum_{i_{m+1}=1}^{N} A_{i_1 i_2 \ldots i_{m+1}}, \\
E_j(v_k) &= \sum_{t=1,\ \text{such that}\ O_t = v_k}^{T-m} \gamma_t(j), \\
\overline{b_j(v_k)} &= E_j(v_k) / \sum_{k=1}^{M} E_j(v_k).
\end{aligned}
\right.
$$

8.2.4 The EM Algorithm

In this subsection, we discuss the convergence of the EM algorithm. We begin with the following lemma.

Lemma 8.1. *Given $p_i, q_i \geq 0$ such that*

$$\sum_i p_i = \sum_i q_i = 1,$$

then

$$\sum_i p_i \log \frac{p_i}{q_i} \geq 0$$

and the equality holds if and only if $p_i = q_i$ for all i.

Proof. Suppose that $p_i, q_i \geq 0$ and

$$\sum_i p_i = \sum_i q_i = 1,$$

then we have

$$-\sum_i p_i \log \frac{p_i}{q_i} = \sum_i p_i \log \frac{q_i}{p_i}$$
$$\leq \sum_i p_i (\frac{q_i}{p_i} - 1)$$
$$= \sum_i (q_i - p_i)$$
$$= 0.$$

This is true because we have the following inequality

$$\log x \leq x - 1 \quad \text{for} \quad x \geq 0$$

and the equality holds if and only if $x = 1$. Hence the result follows.

Now, suppose we have a model with parameter set Λ and we want to obtain a better model with parameter set $\overline{\Lambda}$. Then one can consider the log likelihood as follows:

$$\log P[O|\overline{\Lambda}] = \sum_Q \log P[O, Q|\overline{\Lambda}].$$

Since

$$P[O, Q|\overline{\Lambda}] = P[Q|O, \overline{\Lambda}]P[O|\overline{\Lambda}],$$

we get

$$\log P[O|\overline{\Lambda}] = \log P[O, Q|\overline{\Lambda}] - \log P[Q|O, \overline{\Lambda}].$$

By multiplying this with $P[Q|O, \Lambda]$ and summing over Q, we get the following

$$\log P[O|\overline{A}] = \sum_Q P[Q|O, A] \log P[O, Q|\overline{A}] - \sum_Q P[Q|O, A] \log P[Q|O, \overline{A}].$$

We denote

$$C(A, \overline{A}) = \sum_Q P[Q|O, A] \log P[O, Q|\overline{A}]$$

then we have

$$\log P[O|\overline{A}] - \log P[O|A] = C(A, \overline{A}) - C(A, A)$$
$$+ \sum_Q P[Q|O, A] \log \frac{P[Q|O, A]}{P[Q|O, \overline{A}]}.$$

The last term of the right-hand-side is the relative entropy of $P[Q|O, A]$ relative to $P[Q|O, \overline{A}]$ which is always non-negative by Lemma 8.1.

Hence we have

$$\log P[O|\overline{A}] - \log P[O|A] \geq C(A, \overline{A}) - C(A, A)$$

and equality holds only if

$$\overline{A} = A$$

or if

$$P[Q|O, \overline{A}] = P[Q|O, A]$$

for some other $\overline{A} \neq A$. By choosing

$$\overline{A} = \arg\max_{A'} C(A, A')$$

one can always make the difference non-negative. Thus the likelihood of the new model is greater than or equal to the likelihood of the old model. In fact, if a maximum is reached then $\overline{A} = A$ and the likelihood remains unchanged. Therefore it can be shown that the EM algorithm converges to a (local or global) maximum.

Proposition 8.2. *The EM algorithm converges to a (local or global) maximum.*

8.2.5 Heuristic Method for Higher-order HMMs

The conventional model for an mth order Markov model has $O(N^{m+1})$ unknown parameters (transition probabilities) where N is number of states. The major problem in using such kind of model is that the number of parameters (transition probabilities) increases exponentially with respect to the order of the model. This large number of parameters discourages the use of higher-order Markov models directly. In this subsection, we develop an efficient estimation method for building a higher-order HMM when the observation symbol probability distribution B is known.

We consider the higher-order Markov model discussed in Chapter 6 whose number of states is linear in m. Our idea is to approximate an nth order Markov model of the demand as follows:

$$\mathbf{Q}_{t+m} = \sum_{i=1}^{m} \lambda_i P_i \mathbf{Q}_{t+m-i} \tag{8.2}$$

where $\mathbf{Q}_{t+i}$ is the state probability distribution vector at time $(t + i)$. In this model we assume that $\mathbf{Q}_{t+n+1}$ depends on $\mathbf{Q}_{t+i}$ $(i = 1, 2, \ldots, n)$ via the matrices P_i and the parameters λ_i. One may relate P_i to the ith step transition probability matrix for the hidden states. In the model, the number of states is $O(mN^2)$ whereas the conventional nth order Markov model has $O(N^{m+1})$ parameters to be determined.

Given the hidden state probability distribution, the observation probability distribution is given by

$$\mathbf{Y}_t = B\mathbf{X}_t \tag{8.3}$$

where B is the emission probabilities matrix. Hence (8.2) and (8.3) form a higher-order HMM.

For Model (8.2), in Chapter 6 we have proposed efficient methods to estimate A_i and λ_i. Given an observed sequence of $\{\mathbf{X}_t\}_{t=1}^{T}$, A_i are estimated by first counting the i-step transition frequency from the observed data sequence and then by normalization to get the transition probabilities. In Chapter 6, we have proved that

$$\lim_{t \to \infty} \mathbf{X}_t = \mathbf{Z} \quad \text{and} \quad \mathbf{Z} = \sum_{i=1}^{m} \lambda_i P_i \mathbf{Z}$$

where $\mathbf{Z}$ can be estimated from $\{\mathbf{X}_t\}_{t=1}^{T}$ by first counting the occurrence frequency of each state and then by normalization. They considered solving λ_i by the following minimization problem:

$$\min \left\| \mathbf{Z} - \sum_{i=1}^{m} \lambda_i P_i \mathbf{Z} \right\|$$

subject to

$$\sum_{i=1}^{m} \lambda_i = 1 \quad \text{and} \quad \lambda_i \geq 0.$$

It can be shown easily that if $\|.\|$ is taken to be $\|.\|_1$ or $\|.\|_\infty$ then the above problem can be reduced to a linear programming problem and hence can be solved efficiently.

Consider a higher-order HMM with known emission probabilities B and observation data sequence

$$O_1 O_2 \ldots O_T,$$

how to choose A_i and λ_i so as to build a higher-order HMM? We note that by (8.3), the stationary probability distribution vector for the observation symbols is given by $\mathbf{W} = B\mathbf{Z}$. Therefore if $\mathbf{W}$ can be estimated and B is given, the probability distribution vector $\mathbf{Z}$ for the hidden states can be obtained. For such stationary vector $\mathbf{Z}$, the first-order transition probability matrix A for the hidden states is then given by

$$A = \mathbf{Z}(1, 1, \ldots, 1)^T \tag{8.4}$$

(noting that $A\mathbf{Z} = vecZ$). With this idea, we propose the following steps to construct a higher-order HMM.

Step 1: The lth element of $\mathbf{W}$ is approximated by

$$\frac{1}{T} \sum_{i=1}^{T} I_{O_i = v_l}.$$

Step 2: From (8.3), we expect $(\mathbf{W} - B\mathbf{Z})$ to be close to the zero vector. Therefore we consider solving $\mathbf{Z}$ by minimizing

$$\|\mathbf{W} - B\mathbf{Z}\|_{\infty}.$$

Step 3: Find the most probable hidden sequence Q_1, Q_2, ..., Q_T based on the observation sequence

$$O_1, O_2, \ldots, O_T$$

and the matrix A is computed by (8.4).

Step 4: With the most probable hidden sequence

$$Q_1, Q_2, \ldots, Q_T,$$

we can estimate P_i by counting the number of the transition frequency of the hidden states and then by normalization.

Step 5: Solve λ_i by solving

$$\min \|\mathbf{Z} - \sum_{i=1}^{m} \lambda_i P_i \mathbf{Z}\|_{\infty}$$

subject to

$$\sum_{i=1}^{m} \lambda_i = 1 \quad \text{and} \quad \lambda_i \geq 0.$$

The advantage of our proposed method is that one can solve the model parameters efficiently with reasonable accuracy. In the next section, we illustrate the effectiveness of this efficient method.

8.2.6 Experimental Results

In this section, we test our higher-order HMMs and the heuristic model for the CpG island data. We simulate a higher-order HMM for the CpG islands. In the genome where-ever the dinucleotide CG occurs (frequently written CpG to distinguish it from the C-G base pair across the two strands) the C nucleotide (cytosine) is typically chemically modified by methylation. There is a relatively high chance of this methyl-C mutating into a T, with the consequence that in general CpG dinucleotides are rarer in the genome than would be expected from the independent probabilities of C and G. Usually, this part corresponds to the promoters or "start" regions of many genes [31]. In DNA sequence analysis, we often focus on which part of the sequence belongs to CpG island and which part of the sequence belongs to non-CpG islands. In the HMM formulation, we have two hidden states ($N = 2$):

$$S_1 = \text{CpG island} \quad \text{and} \quad S_2 = \text{non} - \text{CpG island},$$

and we have four observations symbols ($M = 4$):

$$v_1 = A, \quad v_2 = C, \quad v_3 = G, \quad v_4 = T.$$

The model parameters based on the information of CpG island are used. The transition probabilities are then given by

$$P(q_t = S_1|q_{t-1} = S_1, q_{t-2} = S_1) = 0.72,$$
$$P(q_t = S_1|q_{t-1} = S_1, q_{t-2} = S_2) = 0.81,$$
$$P(q_t = S_1|q_{t-1} = S_2, q_{t-2} = S_1) = 0.12,$$
$$P(q_t = S_1|q_{t-1} = S_2, q_{t-2} = S_2) = 0.21,$$
$$P(q_t = S_2|q_{t-1} = S_1, q_{t-2} = S_1) = 0.28,$$
$$P(q_t = S_2|q_{t-1} = S_1, q_{t-2} = S_2) = 0.19,$$
$$P(q_t = S_2|q_{t-1} = S_2, q_{t-2} = S_1) = 0.88,$$
$$P(q_t = S_2|q_{t-1} = S_2, q_{t-2} = S_2) = 0.79.$$

and

$$P(O_t = A|q_t = S_1) = 0.1546,$$
$$P(O_t = C|q_t = S_1) = 0.3412,$$
$$P(O_t = G|q_t = S_1) = 0.3497,$$
$$P(O_t = T|q_t = S_1) = 0.1544,$$
$$P(O_t = A|q_t = S_2) = 0.2619,$$
$$P(O_t = C|q_t = S_2) = 0.2463,$$
$$P(O_t = G|q_t = S_2) = 0.2389,$$
$$P(O_t = T|q_t = S_2) = 0.2529.$$

Given these values, the HMM can be used as a generator to give an observation sequence. We generate 100 observation sequences of length $T = 3000$. Based on these observation sequences, we train three models. The three models assume that the hidden states sequence is a first-order model, a second-order model and a third-order model respectively. We calculate

$$P(O|\Lambda) \quad \text{and} \quad P(Q, O|\Lambda)$$

for each of the models. We also report the results obtained by using our proposed heuristic model. The average results of 100 comparisons are given in Table 8.1. It is clear that the proposed estimation algorithm can recover the second-order Markov model of the hidden states.

Table 8.1. $\log P[O|\Lambda]$.

	First-order	Second-order	Third-order
The Heuristic Method	-1381	-1378	-1381
EM Algorithm (no. of iter)	-1377 (2.7)	-1375 (3.5)	-1377 (3.4)

Finally, we present the computation times (per iteration) required for the heuristic method and the EM algorithms in Table 8.2. We remark that the heuristic method requires only one iteration. we see that the proposed heuristic method is efficient.

Table 8.2. Computational times in seconds.

	First-order	Second-order	Third-order
The Heuristic Method	1.16	1.98	5.05
EM Algorithm	4.02	12.88	40.15

8.3 The Interactive Hidden Markov Model

In this section, we propose an Interactive Hidden Markov Model (IHMM) where the transitions of hidden states depend on the current observable states. The IHHM is a generalization of the HMM discussed in Chapter 4. We note that this kind of HMM is different from classical HMMs where the next hidden states are governed by the previous hidden states only. An example is given to demonstrate IHMM. We then extend the results to give a general IHMM.

8.3.1 An Example

Suppose that we are given a categorical data sequence (in steady state) of volumn of transactions as follows:

$$1, 2, 1, 2, 1, 2, 2, 4, 1, 2, 2, 1, 3, 3, 4, 1.$$

Here 1=high transaction volume, 2= medium transaction volume, 3=low transaction volume and 4=very low transaction volume. Suppose there are two hidden states: A (bull market period) and B (bear market period). In period A, the probability distribution of the transaction volume is assumed to follow

$$(1/4, 1/4, 1/4, 1/4).$$

In period B, the probability distribution of the transaction volume is assumed to follow

$$(1/6, 1/6, 1/3, 1/3).$$

In the proposed model, we assume that hidden states are unobservable but the transaction volume are observable. We would like to uncover the hidden state by modelling the dynamics by a Markov chain.

In the Markov chain, the states are

$$A, B, 1, 2, 3, 4.$$

We assume that when the observable state is i then the probabilities that the hidden state is A and B are given by α_i and $1 - \alpha_i$ (depending on i) respectively in next time step. The transition probability matrix governing the Markov chain is given by

$$P_1 = \begin{pmatrix} 0 & 0 & 1/4 & 1/4 & 1/4 & 1/4 \\ 0 & 0 & 1/6 & 1/6 & 1/3 & 1/3 \\ \alpha_1 & 1-\alpha_1 & 0 & 0 & 0 & 0 \\ \alpha_2 & 1-\alpha_2 & 0 & 0 & 0 & 0 \\ \alpha_3 & 1-\alpha_3 & 0 & 0 & 0 & 0 \\ \alpha_4 & 1-\alpha_4 & 0 & 0 & 0 & 0 \end{pmatrix}.$$

8.3.2 Estimation of Parameters

In order to define the IHMM, one has to estimate the model parameters $\alpha_1, \alpha_2, \alpha_3$ and α_4 from an observed data sequence. One may consider the following two-step transition probability matrix as follows:

$$P_1^2 = \begin{pmatrix} \frac{\alpha_1+\alpha_2+\alpha_3+\alpha_4}{4} & 1 - \frac{\alpha_1+\alpha_2+\alpha_3+\alpha_4}{4} & 0 & 0 & 0 & 0 \\ \frac{\alpha_1+\alpha_2}{6} + \frac{\alpha_3+\alpha_4}{3} & 1 - \frac{\alpha_1+\alpha_2}{6} - \frac{\alpha_3+\alpha_4}{3} & 0 & 0 & 0 & 0 \\ 0 & 0 & \frac{1}{6}+\frac{\alpha_1}{12} & \frac{1}{6}+\frac{\alpha_1}{12} & \frac{1}{3}-\frac{\alpha_1}{12} & \frac{1}{3}-\frac{\alpha_1}{12} \\ 0 & 0 & \frac{1}{6}+\frac{\alpha_2}{12} & \frac{1}{6}+\frac{\alpha_2}{12} & \frac{1}{3}-\frac{\alpha_2}{12} & \frac{1}{3}-\frac{\alpha_2}{12} \\ 0 & 0 & \frac{1}{6}+\frac{\alpha_3}{12} & \frac{1}{6}+\frac{\alpha_3}{12} & \frac{1}{3}-\frac{\alpha_3}{12} & \frac{1}{3}-\frac{\alpha_3}{12} \\ 0 & 0 & \frac{1}{6}+\frac{\alpha_4}{12} & \frac{1}{6}+\frac{\alpha_4}{12} & \frac{1}{3}-\frac{\alpha_4}{12} & \frac{1}{3}-\frac{\alpha_4}{12} \end{pmatrix}.$$

Using the same track as in Chapter 4, one can extract the one-step transition probability matrix of the observable states from P_2^2 as follows:

$$\tilde{P}_2 = \begin{pmatrix} \frac{1}{6} + \frac{\alpha_1}{12} & \frac{1}{6} + \frac{\alpha_1}{12} & \frac{1}{3} - \frac{\alpha_1}{12} & \frac{1}{3} - \frac{\alpha_1}{12} \\ \frac{1}{6} + \frac{\alpha_2}{12} & \frac{1}{6} + \frac{\alpha_2}{12} & \frac{1}{3} - \frac{\alpha_2}{12} & \frac{1}{3} - \frac{\alpha_2}{12} \\ \frac{1}{6} + \frac{\alpha_3}{12} & \frac{1}{6} + \frac{\alpha_3}{12} & \frac{1}{3} - \frac{\alpha_3}{12} & \frac{1}{3} - \frac{\alpha_3}{12} \\ \frac{1}{6} + \frac{\alpha_4}{12} & \frac{1}{6} + \frac{\alpha_4}{12} & \frac{1}{3} - \frac{\alpha_4}{12} & \frac{1}{3} - \frac{\alpha_4}{12} \end{pmatrix}.$$

However, in this case, we do not have a closed form solution for the stationary distribution of the process. To estimate the parameter α_i, we first estimate the one-step transition probability matrix from the observed sequence. This can be done by counting the transition frequencies of the states in the observed sequence and we have

$$\hat{P}_2 = \begin{pmatrix} 0 & \frac{4}{5} & \frac{1}{5} & 0 \\ \frac{1}{2} & \frac{1}{3} & 0 & \frac{1}{6} \\ 0 & 0 & \frac{1}{2} & \frac{1}{2} \\ 1 & 0 & 0 & 0 \end{pmatrix}.$$

We expect that

$$\tilde{P}_2 \approx \hat{P}_2$$

and hence α_i can be obtained by solving the following minimization problem:

$$\min_{\alpha_i} ||\tilde{P}_2 - \hat{P}_2||_F^2 \qquad (8.5)$$

subject to

$$0 \leq \alpha_i \leq 1.$$

Here $||.||_F$ is the Frobenius norm, i.e.

$$||A||_F^2 = \sum_{i=1}^{n} \sum_{i=1}^{n} A_{ij}^2.$$

This is equivalent to solve the following four independent minimization problems (i) - (iv) and they can be solved in parallel. This is an advantage of the estimation method. We remark that one can also consider other matrix norms for the objective function (8.5), let us say $||.||_{M_1}$ or $||.||_{M_\infty}$ and they may result in linear programming problems.

(i) α_1: $\displaystyle\min_{0 \leq \alpha_1 \leq 1}\{(\frac{1}{6} + \frac{\alpha_1}{12})^2 + (\frac{1}{6} + \frac{\alpha_1}{12} - \frac{4}{5})^2 + (\frac{1}{3} - \frac{\alpha_1}{12} - \frac{1}{5})^2 + (\frac{1}{3} - \frac{\alpha_1}{12})^2\};$

(ii) α_2: $\displaystyle\min_{0 \leq \alpha_2 \leq 1}\{(\frac{1}{6} + \frac{\alpha_1}{12} - \frac{1}{2})^2 + (\frac{1}{6} + \frac{\alpha_1}{12} - \frac{1}{3})^2 + (\frac{1}{3} - \frac{\alpha_1}{12})^2 + (\frac{1}{3} - \frac{\alpha_1}{12} - \frac{1}{6})^2\};$

(iii) α_3: $\displaystyle\min_{0 \leq \alpha_3 \leq 1}\{(\frac{1}{6} + \frac{\alpha_1}{12})^2 + (\frac{1}{6} + \frac{\alpha_1}{12})^2 + (\frac{1}{3} - \frac{\alpha_1}{12} - \frac{1}{2})^2 + (\frac{1}{3} - \frac{\alpha_1}{12} - \frac{1}{2})^2\};$

(iv) α_4: $\min\limits_{0 \le \alpha_4 \le 1}\{(\frac{1}{6}+\frac{\alpha_1}{12}-1)^2+(\frac{1}{6}+\frac{\alpha_1}{12})^2+(\frac{1}{3}-\frac{\alpha_1}{12})^2+(\frac{1}{3}-\frac{\alpha_1}{12})^2\}.$

Solving the above optimization problems, we have

$$\alpha_1^* = 1, \quad \alpha_2^* = 1, \quad \alpha_3^* = 0, \quad \alpha_4^* = 1.$$

Hence we have

$$P_2 = \left(\begin{array}{cc|cccc} 0 & 0 & 1/4 & 1/4 & 1/4 & 1/4 \\ 0 & 0 & 1/6 & 1/6 & 1/3 & 1/3 \\ \hline 1 & 0 & 0 & 0 & 0 & 0 \\ 1 & 0 & 0 & 0 & 0 & 0 \\ 0 & 1 & 0 & 0 & 0 & 0 \\ 1 & 0 & 0 & 0 & 0 & 0 \end{array}\right) \tag{8.6}$$

and

$$P_2^2 = \left(\begin{array}{cc|cccc} 3/4 & 1/4 & 0 & 0 & 0 & 0 \\ 2/3 & 1/3 & 0 & 0 & 0 & 0 \\ \hline 0 & 0 & 1/4 & 1/4 & 1/4 & 1/4 \\ 0 & 0 & 1/4 & 1/4 & 1/4 & 1/4 \\ 0 & 0 & 1/6 & 1/6 & 1/3 & 1/3 \\ 0 & 0 & 1/4 & 1/4 & 1/4 & 1/4 \end{array}\right). \tag{8.7}$$

8.3.3 Extension to the General Case

The method can be extended to a general case of m hidden states and n observable states. We note the one-step transition probability matrix of the observable states is given by

$$\tilde{P}_2 = \begin{pmatrix} \alpha_{11} & \alpha_{12} & \cdots & \alpha_{1m} \\ \alpha_{21} & \alpha_{22} & \cdots & \alpha_{2m} \\ \vdots & \vdots & \vdots & \vdots \\ \alpha_{n1} & \alpha_{m2} & \cdots & \alpha_{nm} \end{pmatrix} \begin{pmatrix} p_{11} & p_{12} & \cdots & p_{1n} \\ p_{21} & p_{22} & \cdots & p_{2n} \\ \vdots & \vdots & \vdots & \vdots \\ p_{m1} & p_{m2} & \cdots & p_{mn} \end{pmatrix}, \tag{8.8}$$

i.e.

$$[\tilde{P}_2]_{ij} = \sum_{k=1}^{m} \alpha_{ik}p_{kj} \quad i,j = 1,2,\ldots,n.$$

Here we assume that α_{ij} are unknowns and the probabilities p_{ij} are given. Suppose $[Q]_{ij}$ is the one-step transition probability matrix estimated from the observed sequence. Then for each fixed i, $\alpha_{ij}, j = 1,2,\ldots,m$ can be obtained by solving the following constrained least squares problem:

$$\min_{\alpha_{ik}}\left\{\sum_{j=1}^{n}\left(\sum_{k=1}^{m}\alpha_{ik}p_{kj}-[Q]_{ij}\right)^2\right\}$$

subject to

$$\sum_{k=1}^{m} \alpha_{ik} = 1$$

and

$$\alpha_{ik} \geq 0 \quad \text{for all } i, k.$$

The idea of the IHMM presented in this subsection is further extended to address the following applications and problems in Ching et al. [67].

(i) IHMM is applied to some practical data sequences in sales demand data sequences.
(ii) there are only a few works on modelling the non-linear behavior of categorical time series can be found in literature. In the continuous-state case, the threshold auto-regressive model is a well-known approach. The idea is to provide a piecewise linear approximation to a non-linear autoregressive time series model by dividing the state space into several regimes via threshold principle. The IHMM provides a first-order approximation of the non-linear behavior of categorical time series by dividing the state space of the Markov chain process into several regimes.

8.4 The Double Higher-order Hidden Markov Model

In this section, we present a discrete model for extracting information about the hidden or unobservable states information from two observation sequences. The observations in each sequence not only depends on the hidden state information, but also depends on its previous observations. It is clear that both the dynamics of hidden states and observation states are required to model higher-order Markov chains. We call this kind of models to be Double Higher-order Hidden Markov Models (DHHMMs).

The model can be described as follows. We write $\mathcal{T}$ for the time index set

$$\{0, 1, 2, \ldots\}$$

of the model. Let $\{V_t\}_{t \in \mathcal{T}}$ be an unobservable process representing the hidden states over different time periods. We assume that $\{V_t\}_{t \in \mathcal{T}}$ is an nth-order discrete-time time-homogeneous Markov chain process with the state space

$$\mathcal{V} = \{v_1, v_2, \ldots, v_M\}.$$

The state transition probabilities matrix

$$A = \{a(j_{t+n})\}$$

of the nth-order Markov chain $\{V_t\}_{t \in \mathcal{T}}$ are given by

$$a(j_{t+n}) = P(V_{t+n} = v_{j_{t+n}} | V_t = v_{j_t}, \ldots, V_{t+n-1} = v_{j_{t+n-1}})$$
$$1 \leq j_t, \ldots, j_{t+n-1} \leq M. \qquad (2.1)$$

To determine the probability structure for the nth-order Markov chain $\{V_t\}_{t \in \mathcal{T}}$ uniquely, we need to specify the initial state conditional probabilities

$$\Pi = \{\pi(i_j)\}$$

as follows:

$$\pi(j_k) = P(V_k = v_{j_k} | V_1 = v_{j_1}, V_2 = v_{j_2}, \ldots, V_{k-1} = v_{j_{k-1}}), \quad 1 \leq k \leq n.$$
$$(2.2)$$

Let
$$\{I_t\}_{t \in \mathcal{T}}$$

for a stochastic process and it is assumed to be a (l, n)-order double hidden Markov chain process. Their corresponding states are given by

$$\{i_t\}_{t \in \mathcal{T}}.$$

Let
$$\mathbf{I}_t = (I_t, I_{t-1}, \ldots, I_{t-l+1})$$

and
$$\mathbf{i}_t = (i_t, i_{t-1}, \ldots, i_{t-l+1}).$$

Then, we assume that the transition probabilities matrix

$$B = \{b_{\mathbf{i}_t, v}(i_{t+1})\}$$

of the process $\{I_t\}_{t \in \mathcal{T}}$ when $\mathbf{I}_t = \mathbf{i}_t$ and the hidden state $V_{t+1} = v$. The initial distribution Π for $\{I_t\}_{t \in \mathcal{T}}$ should be specified. Given appropriate values for n, M, I, A, l, Π and B, the DHHMM can be adopted to describe the generator that drives the realization of the observable sequence

$$I = I_1 I_2 \ldots I_T,$$

where T is the number of observations in the sequence. In order to determine the DHHMM for our applications one can apply similar method of maximum likelihood estimation and the EM algorithm discussed in Section 8.2. A detailed discussion of the model and method of estimation with applications to the extraction of unobservable states of an economy from observable spot interest rates and credit ratings can be found in Siu et al. [189].

8.5 Summary

In this chapter, we present several new frameworks of hidden Markov models (HMMs). They include Higher-order Hidden Markov Model (HHMM), Interactive Hidden Markov Model (IHMM) and Double Higher-order Hidden Markov Model (DHHMM). For both HHMM and IHMM, we present both methods and efficient algorithms for the estimation of model parameters. Further research can be done in the applications of these new HMMs.

References

1. Albrecht D, Zukerman I and Nicholson A (1999) Pre-sending Documents on the WWW: A Comparative Study, Proceedings of the Sixteenth International Joint Conference on Artificial Intelligence IJCAI99.
2. Adke S and Deshmukh D (1988) Limit Distribution of a High Order Markov Chain, Journal of Royal Statistical Society, Series B, 50:105–108.
3. Akutsu T, Miyano S and Kuhara S (2000) Inferring Qualitative Relations in Genetic Networks and Metabolic Arrays, Bioinformatics, 16:727–734.
4. Altman E (1999) Constrained Markov Decision Processes, Chapman and Hall/CRC.
5. Ammar G and Gragg W (1988) Superfast Solution of Real Positive Definite Toeplitz Systems, SIAM Journal of Matrix Analysis and Its Applications, 9:61–76.
6. Artzner P and Delbaen F (1997) Default Risk Premium and Incomplete Markets, Mathematical Finance, 5:187–195.
7. Artzner P, Delbaen F, Eber J and Heath D (1997) Thinking Coherently, Risk, 10:68–71.
8. Avery P (1987) The Analysis of Intron Data and Their Use in the Detection of Short Signals, Journal of Molecular Evolution, 26:335–340.
9. Avrachenkov L and Litvak N (2004) Decomposition of the Google PageRank and Optimal Linking Strategy, Research Report, INRIA, Sophia Antipolis.
10. Axsäter S (1990) Modelling Emergency Lateral Transshipments in Inventory Systems, Management Science, 36:1329–1338.
11. Axelsson O (1996) Iterative Solution Methods, Cambridge University Press, N.Y.
12. Baldi P, Frasconi P and Smith P (2003) Modeling the Internet and the Web, Wiley, England.
13. Bandholz H and Funke M (2003) In Search of Leading Indicators of Economic Activity in Germany, Journal of Forecasting, 22:277–297.
14. Baum L (1972) An Inequality and Associated Maximization Techniques in statistical Estimation for Probabilistic Function of Markov Processes, Inequality, 3:1–8.
15. Bell D, Atkinson J and Carlson J (1999) Centrality Measures for Disease Transmission Networks, Social Networks, 21:1–21.

16. Berman A and Plemmons R (1994) Nonnegative matrices in the Mathematical Sciences, Society for Industrial and Applied Mathematics, Philadelphia.

17. Bernardo J and Smith A (2001) Bayesian Theory, John Wiley & Sons, New York.

18. Berger P and Nasr N (1998) Customer Lifetime Value: Marketing Models and Applications, Journal of Interactive Marketing, 12:17–30.

19. Berger P and Nasr N (2001) The Allocation of Promotion Budget to Maximize Customer Equity, Omega, 29:49–61.

20. Best P (1998) Implementing Value at Risk, John Wiley & Sons, England.

21. Bini D, Latouche G and Meini B (2005) Numerical Methods for Structured Markov Chains Oxford University Press, New York.

22. Blattberg R and Deighton J (1996) Manage Market by the Customer Equity, Harvard Business Review, 73:136–144.

23. Blumberg D (2005) Introduction to Management of Reverse Logistics and Closed Loop Supply Chain Processes CRC Press, Boca Raton.

24. Blattner F, Plunkett G, Boch C, Perna N, Burland V, Riley M, Collado-Vides J, Glasner J, Rode C, Mayhew G, Gregor J, Davis N, Kirkpatrick H, Goeden M, Rose D, Mau B and Shao Y (1997) The Complete Genome Sequence of *Escherichia coli K − 12*, Science 227:1453–1462.

25. Bonacich P and Lloyd P (2001) Eigenvector-like Measures of Centrality for Asymmetric Relations, Social Networks, 23:191–201.

26. Bonacich P and Lloyd P (2004) Calculating Status with Negative Relations, Social Networks, 26:331–338.

27. Bodnar J (1997) Programming the Drosophila Embryo. Journal of Theoretical Biology, 188:391–445.

28. Borodovskii M, Sprizhitskii A, Golovanov I and Aleksandrov A (1986) Statistical Patterns in Primary Structures of the Functional Regions of Genome in *Escherichia coli-*, Molecular Biology, 20:826–833.

29. Bower J (2001) Computational Moeling of Genetic and Biochemical Networks, MIT Press, Cambridge, M.A.

30. Boyle P, Siu T and Yang H (2002) Risk and Probability Measures, Risk, 15(7):53–57.

31. Bird A (1987) CpG Islands as Gene Markers in the Vertebrate Nucleus, Trends in Genetics, 3:342–347.

32. Bramble J (1993) Multigrid Methods, Longman Scientific and Technical, Essex, England.

33. Brockwell P and Davis R (1991) Time Series: Theory and Methods, Springer-Verlag, New York.

34. Buchholz P. (1994) A class of Hierarchical Queueing Networks and their Analysis, Queueing Systems, 15:59–80.

35. Buchholz P. (1995) Hierarchical Markovian Models: Symmetries and Aggregation, Performance Evaluation, 22:93–110.

36. Buchholz P. (1995) Equivalence Relations for Stochastic Automata Networks. Computations of Markov chains: Proceedings of the 2nd international workshop On numerical solutions of Markov chains. Kluwer, 197–216.

37. Bühlmann H (1967) Experience Rating and Credibility Theory, ASTIN Bulletin, 4:199–207.

38. Bunch J (1985) Stability of Methods for Solving Toeplitz Systems of Equations, SIAM Journal of Scientific and Statistical Computing, 6:349–364.

39. Bunke H and Caelli T (2001) Hidden Markov models : applications in computer vision, Editors, Horst Bunke, Terry Caelli, Singapore, World Scientific.

40. Buzacott J and Shanthikumar J (1993) Stochastic Models of Manufacturing Systems, Prentice-Hall International Editions, New Jersey.

41. Camba-Mendaz G, Smith R, Kapetanios G and Weale M (2001) An Automatic Leading Indicator of Economic Activity: Forecasting GDP Growth for European Countries, Econometrics Journal, 4:556–590.

42. Carpenter P (1995) Customer Lifetime Value: Do the Math., Marketing Computers, 15:18–19.

43. Chan R and Ching W (1996) Toeplitz-circulant Preconditioners for Toeplitz Systems and Their Applications to Queueing Networks with Batch Arrivals, SIAM Journal of Scientific Computing, 17:762–772.

44. Chan R and Ching W (2000) Circulant Preconditioners for Stochastic Automata Networks, Numerise Mathematik, 87:35–57.

45. Chan R, Ma K and Ching W (2005) Boundary Value Methods for Solving Transient Solutions of Markovian Queueing Networks, Journal of Applied Mathematics and Computations, to appear.

46. Chan R and Ng M (1996) Conjugate Gradient Method for Toeplitz Systems, SIAM Reviews, 38:427–482.

47. Chang Q, Ma S and Lei G (1999) Algebraic Multigrid Method for Queueing Networks. International Journal of Computational Mathematics, 70:539–552.

48. Ching W (1997) Circulant Preconditioners for Failure Prone Manufacturing Systems, Linear Algebra and Its Applications, 266:161–180.

49. Ching W (1997) Markov Modulated Poisson Processes for Multi-location Inventory Problems, International Journal of Production Economics, 53:217–223.

50. Ching W (1998) Iterative Methods for Manufacturing Systems of Two Stations in Tandem, Applied Mathematics Letters, 11:7–12.

51. Ching W (2001) Machine Repairing Models for Production Systems, International Journal of Production Economics, 70:257–266.

52. Ching W (2001) Iterative Methods for Queuing and Manufacturing Systems, Springer Monographs in Mathematics, Springer, London.

53. Ching W (2001) Markovian Approximation for Manufacturing Systems of Unreliable Machines in Tandem, International Journal of Naval Research Logistics, 48:65–78.

54. Ching W (2003) Iterative Methods for Queuing Systems with Batch Arrivals and Negative Customers, BIT 43:285–296.

55. Ching W, Chan R and Zhou X (1997) Circulant Preconditioners for Markov Modulated Poisson Processes and Their Applications to Manufacturing Systems, SIAM Journal of Matrix Analysis and Its Applications, 18:464–481.

56. Ching W, Fung E and Ng M (2002) A Multivariate Markov Chain Model for Categorical Data Sequences and Its Applications in Demand Predictions, IMA Journal of Management Mathematics, 13:187–199.

57. Ching W, Fung E and Ng M (2003) A Higher-order Markov Model for the Newsboy's Problem, Journal of Operational Research Society, 54:291–298.

58. Ching W and Loh A (2003) Iterative Methods for Flexible Manufacturing Systems, Journal of Applied Mathematics and Computation, 141:553–564.

59. Ching W and Ng M (2003) Recent Advance in Data Mining and Modeling, World Scientific, Singapore.

194   References

60. Ching W and Ng M. (2004) Building Simple Hidden Markov Models, International Journal of Mathematical Education in Science and Engineering, 35:295–299.
61. Ching W, Ng M and Fung E (2003) Higher-order Hidden Markov Models with Applications to DNA Sequences, IDEAL2003, Lecture Notes in Computer Science, (Liu J, Cheung Y and Yin H (Eds.)) 2690:535–539, Springer.
62. Ching W, Fung E and Ng M (2004) Higher-order Markov Chain Models for Categorical Data Sequences, International Journal of Naval Research Logistics, 51:557–574.
63. Ching W, Fung E and Ng M (2004) Building Higher-order Markov Chain Models with EXCEL, International Journal of Mathematical Education in Science and Technology, 35:921–932.
64. Ching W, Fung E and Ng M (2004) Building Genetic Networks in Gene Expression Patterns, IDEAL2004, Lecture Notes in Computer Science, (Yang Z, Everson R and Yin H (Eds.)) 3177:17–24, Springer.
65. Ching W, Fung E and Ng M (2005) Higher-order Multivariate Markov Chains: Models, Algorithms and Applications, Working paper.
66. Ching W, Fung E, Ng M and Ng T (2003) Multivariate Markov Models for the Correlation of Multiple Biological Sequences International Workshop on Bioinformatics, PAKDD Seoul, Korea, 23–34.
67. Ching W, Ng M, Fung E and Siu T (2005) An Interactive Hidden Markov Model for Categorical Data Sequences, Working paper.
68. Ching W, Ng M and So M (2004) Customer Migration, Campaign Budgeting, Revenue Estimation: The Elasticity of Markov Decision Process on Customer Lifetime Value, Electronic International Journal of Advanced Modeling and Optimization, 6(2):65–80.
69. Ching W, Ng M and Wong K (2004) Hidden Markov Models and Its Applications to Customer Relationship Management, IMA Journal of Management Mathematics, 15:13–24.
70. Ching W, Ng M, Wong K and Atlman E (2004) Customer Lifetime Value: A Stochastic Programming Approach, Journal of Operational Research Society, 55:860–868.
71. Ching W, Ng M and Zhang S (2005) On Computation with Higher-order Markov Chain, Current Trends in High Performance Computing and Its Applications Proceedings of the International Conference on High Performance Computing and Applications, August 8-10, 2004, Shanghai, China (Zhang W, Chen Z, Glowinski R, and Tong W (Eds.)) 15–24, Springer.
72. Ching W, Ng M and Wong K (2003) Higher-order Markov Decision Process and Its Applications in Customer Lifetime Values, The 32nd International Conference on Computers and Industrial Engineering, Limerick, Ireland 2: 821–826.
73. Ching W, Ng M and Yuen W (2003) A Direct Method for Block-Toeplitz Systems with Applications to Re-Manufacturing Systems, Lecture Notes in Computer Science 2667, (Kumar V, Gavrilova M, Tan C and L'Ecuyer P (Eds.)) 1:912–920, Springer.
74. Ching W, Yuen W, Ng M and Zhang S (2005) A Linear Programming Approach for Solving Optimal Advertising Policy, IMA Journal of Management Mathematics, to appear.
75. Ching W and Yuen W (2002) Iterative Methods for Re-manufacturing Systems, International Journal of Applied Mathematics, 9:335–347.

76. Ching W, Yuen W and Loh A (2003) An Inventory Model with Returns and Lateral Transshipments, Journal of Operational Research Society, 54:636–641.

77. Ching W, Ng M and Yuen W (2005), A Direct Method for Solving Block-Toeplitz with Near-Circulant-Block Systems with Applications to Hybrid Manufacturing Systems, Journal of Numerical Linear Algebra with Applications, to appear.

78. Cho D and Parlar M (1991) A Survey of Maintenance Models for Multi-unit Systems, European Journal of Operational Research, 51:1–23.

79. Chvatal V (1983) Linear Programming, Freeman, New York.

80. Cooper R (1972) Introduction to Queueing Theory, Macmillan, New York.

81. Datta A, Bittner M and Dougherty E (2003) External Control in Markovian Genetic Regulatory Networks, Machine Learning, 52:169–191.

82. Davis P (1979) Circulant Matrices, John Wiley and Sons, New York.

83. de Jong H (2002) Modeling and Simulation of Genetic Regulatory Systems: A Literature Review, Journal of Computational. Biology, 9:69–103.

84. Dekker R, Fleischmann M, Inderfurth K and van Wassenhove L (2004) Reverse Logistics : Quantitative Models for Closed-loop Supply Chains Springer, Berlin.

85. Dowd K (1998) Beyond Value at Risk: The Science of Risk Management, John Wiley & Sons , New York.

86. Duffie D and Pan J (1997) An Overview of Value at Risk. Journal of Derivatives, 4(3):7–49.

87. Duffie D and Pan J (2001) Analytical Value-at-risk with Jumps and Credit Risk, Finance and Stochastic, 5(2):155–180.

88. Duffie D, Schroder M and Skiadas C (1996) Recursive Valuation of Defaultable Securities and the Timing of the Resolution of Uncertainty, Annal of Applied Probability, 6:1075–1090.

89. DuWors R and Haines G (1990) Event History Analysis Measure of Brand Loyalty, Journal of Marketing Research, 27:485–493.

90. Embrechts P, Mcneil A and Straumann D (1999) Correlation and Dependence in Risk Management: Properties and Pitfalls, Risk, May:69–71.

91. Fang S and Puthenpura S (1993) Linear Optimization and Extensions, Prentice-Hall, New Jersey.

92. Fleischmann M (2001) Quantitative Models for Reverse Logistics, Lecture Notes in Economics and Mathematical Systems, 501, Springer, Berlin.

93. Frey R and McNeil A (2002) VaR and Expected Shortfall in Portfolios of Dependent Credit Risks: Conceptual and Practical Insights, Journal of Banking and Finance, 26:1317–1334.

94. Gelenbe E (1989) Random Neural Networks with Positive and Negative Signals and Product Solution, Neural Computation, 1:501-510.

95. Gelenbe E, Glynn P and Sigman K (1991) Queues with Negative Arrivals, Journal of Applied Probability, 28:245-250.

96. Gelenbe E (1991) Product Form Networks with Negative and Positive Customers, Journal of Applied Probability, 28:656-663.

97. Goldberg D (1989) Genetic Algorithm in Search, Optimization, and Machine Learning, Addison-Wesley.

98. Garfield E (1955) Citation Indexes for Science: A New Dimension in Documentation Through Association of Ideas, Science, 122:108–111.

99. Garfield E (1972) Citation Analysis as a Tool in Journal Evaluation, Science, 178:471–479.

100. Salzberg S, Delcher S, Kasif S and White O (1998) Microbial gene identification using interpolated Markov models, Nuclei Acids Research, 26:544–548.
101. Golub G and van Loan C (1989) Matrix Computations, The John Hopkins University Press, Baltimore.
102. Gowda K and Diday E (1991) Symbolic Clustering Using a New Dissimilarity Measure, Pattern Recognition, 24(6):567–578.
103. Häggström (2002) Finite Markov Chains and Algorithmic Applications, London Mathematical Society, Student Texts 52, Cambridge University Press, Cambridge, U.K.
104. Hall M and Peters G (1996) Genetic Alterations of Cyclins, Cyclin-dependent Kinases, and Cdk Inhibitors in Human Cancer. Advances in Cancer Research, 68:67–108.
105. Hartwell L and Kastan M (1994) Cell Cycle Control and Cancer. Science, 266:1821–1828.
106. Haveliwala T and Kamvar S (2003) The Second Eigenvalue of the Google Matrix, Stanford University, Technical Report.
107. He J, Xu J and Yao X (2000) Solving Equations by Hybrid Evolutionary Computation Techniques, IEEE Transaction on Evoluationary Computations, 4:295–304.
108. Hénaut A and Danchin A (1996) Analysis and Predictions from *Escherichia Coli* Sequences, or *E. coli* In Silico, Escherichia coli and Salmonella, Cellular and Molecular Biology, 1:2047–2065.
109. Hestenes M and Stiefel E (1952) Methods of Conjugate Gradients for Solving Linear Systems, Journal of research of the National Bureau of Standards, 49:490–436.
110. Heyman D (1977) Optimal Disposal Policies for Single-item Inventory System with Returns, Naval Research and Logistics, 24:385–405.
111. Holmes J (1988) Speech synthesis and Recognition, Van Nostrand Reinhold, U.K.
112. Horn R and Johnson C (1985) Matrix analysis, Cambridge University Press.
113. Hu Y, Kiesel R and Perraudin W (2002) The Estimation of Transition Matrices for Sovereign Ratings, Journal of Banking and Finance, 26(7):1383–1406.
114. Huang J, Ng M, Ching W, Cheung D, Ng J (2001) A Cube Model for Web Access Sessions and Cluster Analysis, WEBKDD 2001, Workshop on Mining Web Log Data Across All Customer Touch Points, The Seventh ACM SIGKDD International Conference on Knowledge Discovery and Data Mining, Lecture Notes in Computer Science, (Kohavi R, Masand B, Spiliopoulou M and Srivastava J (Eds.)) 47–58, Springer.
115. Hughes A and Wang P (1995) Media Selection for Database Marketers, Journal of Direct Marketing, 9:79–84.
116. Huang S and Ingber D (2000) Shape-dependent Control of Cell Growth, Differentiation, and Apoptosis: Switching Between Attractors in Cell Regulatory Networks, Experimental Cell Research, 261:91–103.
117. Inderfurth K and van der Laan E (2001) Leadtime Effects and Policy Improvement for Stochastic Inventory Control with Remanufacturing, International Journal of Production Economics, 71:381–390.
118. Jackson B (1985) Winning and Keeping Industrial Customers, Lexington, MA: Lexington Books.
119. Jarrow R and Turnbull S (1995) Pricing Options on Financial Derivatives Subject to Default Risk, Journal of Finance, 50:53–86.

120. Jarrow R, Lando D and Turnbull S (1997) A Markov Model for the Term Structure of Credit Spreads, Review of Financial Studies, 10:481–523.
121. Joachims T, Freitag D and Mitchell T (1997) WebWatch: A Tour Guide for the World Wide Web, Proceedings of the Fifteenth International Joint Conference on Artificial Intelligence IJCAI 97, 770–775.
122. Jorion P (2001) Value at Risk: the New Benchmark for Controlling Market Risk, McGraw-Hill, United States.
123. Kamvar S, Haveliwala T and Golub G (2004) Adaptive Methods for the Computation of PageRank, Linear Algebra and Its Applications, 386:51–65.
124. Kahan W (1958) Gauss-Seidel Methods of Solving Large Systems of Linear Equations. Ph.D. thesis, Toronto, Canada, University of Toronto.
125. Kauffman S (1969) Metabolic Stability and Epigenesis in Randomly Constructed Gene Nets, Journal of Theoretical Biology, 22:437–467.
126. Kauffman S (1969) Homeostasis and Differentiation in Random Genetic Control Networks, Nature, 224:177–178.
127. Kiesmüller G and van der Laan E (2001) An Inventory Model with Dependent Product Demands and Returns International Journal of Production Economics, 72:73–87.
128. Kijima M, Komoribayashi K and Suzuki E (2002) A Multivariate Markov Model for Simulating Correlated Defaults. Journal of Risk, 4:1–32.
129. Kim S, Dougherty E, Chen Y, Sivakumar K, Meltzer P, Trent J and Bittner M (2000) Multivariate Measurement of Gene Expression Relationships, Genomics, 67:201–209.
130. Kincaid D and Cheney W (2002) Numerical Analysis: Mathematics of Scientific Computing, 3rd Edition, Books/Cole Thomson Learning, CA.
131. Kleffe J and Borodovsky M (1992) First and Second Moment of Counts of Words in Random Texts Generated by Markov Chains, CABIO, 8:433–441.
132. Klose A, Speranze G and N. Van Wassenhove L (2002) Quantitative Approaches to Distribution Logistics and Supply Chain Management, Springer, Berlin.
133. Klugman S, Panjer H and Willmot G (1997) Loss Models: From Data to Decisions, John Wiley & Sons, New York.
134. Kotler P and Armstrong G (1995) Principle of Marketing, 7th Edition, Prentice Hall, N.J.
135. Koski T (2001) Hidden Markov Models for Bioinformatics, Kluwer Academic Publisher, Dordrecht.
136. Kaufman L (1982) Matrix Methods for Queueing Problems, SIAM Journal on Scientific and Statistical Computing, 4:525–552.
137. Langville A and Meyer C (2005) A Survey of Eigenvector Methods for Web Information Retrieval SIAM Reviews, 47:135–161.
138. Latouche G and Ramaswami V (1999) Introduction to Matrix Analytic Methods in Stochastic Modeling, SIAM, Philadelphia.
139. Lee P (1997) Bayesian Statistics: An Introduction. Edward Arnold, London.
140. Li W and Kwok M (1989) Some Results on the Estimation of a Higher Order Markov Chain, Department of Statistics, The University of Hong Kong.
141. Lieberman H (1995) Letizia: An Agent that Assists Web Browsing, Proceedings of the Fourteenth International Joint Conference on Artificial Intelligence IJCAI 95, 924–929.
142. Latouche G and Ramaswami V (1999) Introduction to Matrix Analytic Methods in Stochastic Modeling, SIAM, Pennsylvania.

143. Latouche G and Taylor P (2002) Matrix-Analytic Methods Theory and Applications, World Scientific, Singapore.

144. Leonard K (1975) Queueing Systems, Wiley, New York.

145. Lim J (1990) Two-Dimensional Signal and Image Processing, Prentice Hall.

146. Lilien L, Kotler P and Moorthy K (1992) Marketing Models, Prentice Hall, New Jersey.

147. Logan J (1981) A Structural Model of the Higher-order Markov Process Incorporating Reversion Effects, Journal of Mathematical Sociology, 8: 75–89.

148. Lu L, Ching W and Ng M (2004) Exact Algorithms for Singular Tridiagonal Systems with Applications to Markov Chains, Journal of Applied Mathematics and Computation, 159:275–289.

149. MacDonald I and Zucchini W (1997) Hidden Markov and Other Models for Discrete-valued Time Series, Chapman & Hall, London.

150. Mesak H and Means T (1998) Modelling Advertising Budgeting and Allocation Decisions Using Modified Multinomial Logit Market Share Models, Journal of Operational Research Society, 49:1260–1269.

151. Mesak H and Calloway J (1999) Hybrid Subgames and Copycat Games in a Pulsing Model of Advertising Competition, Journal of Operational Research Society, 50:837-849.

152. Mesak H and Zhang H (2001) Optimal Advertising Pulsation Policies: A Dynamic Programming Approach, Journal of Operational Research Society, 11:1244-1255.

153. Mesak H (2003) On Deriving and Validating Comparative Statics of a Symmetric Model of Advertising Competition, Computers and Operations Research, 30:1791-1806.

154. Mendoza L, Thieffry D and Alvarez-Buylla E (1999) Genetic Control of Flower Morphogenesis in Arabidopsis Thaliana: A Logical Analysis, Bioinformatics, 15:593–606.

155. Mowbray A (1914) How Extensive a Payroll Exposure is Necessary to give a Dependent Pure Premium, Proceedings of the Causality Actuarial Society, 1:24–30.

156. Muckstadt J and Isaac M (1981) An Analysis of Single Item Inventory Systems with Returns, International Journal of Naval Research and logistics, 28:237–254.

157. Muckstadt J (2005) Analysis and Algorithms for Service Parts Supply Chains Springer, New York.

158. Nahmias S (1981) Managing Repairable Item Inventory Systems: A Review in TIMS Studies, Management Science 16:253–277.

159. Neuts M (1981) Matrix-geometric Solutions in Stochastic Models : An Algorithmic Approach, Johns Hopkins University Press.

160. Neuts M (1995) Algorithmic Probability : A Collection of Problems, Chapman & Hall, London.

161. Nickell P, Perraudin W and Varotto S (2000) Stability of Rating Transitions, Journal of Banking and Finance, 24(1/2):203–228.

162. Nir F, Michal L, Iftach N and Dana P (2000) Using Bayesian Networks to Analyze Expression Data. Journal of Computational Biology, 7(3-4):601–620.

163. McCormick S (1987) Multigrid Methodst, Society for Industrial and Applied Mathematics, Philadelphia, Pa.

164. Ong M (1999) Internal Credit Risk Models: Capital Allocation and Performance Measurement, Risk Books, London.

165. Ott S, Imoto S and Miyano S (2004) Finding Optimal Models for Small Gene Networks, Pacific Symposium on Biocomputing, 9:557–567.
166. Page L, Brin S, Motwani R and Winograd T (1998) The PageRank Citation Ranking: Bring Order to the Web, Technical Report, Stanford University.
167. Patton A (2004) Modelling Asymmetric Exchange Rate Dependence, Working Paper, London School of Economics, United Kingdom.
168. Penza P and Bansal V (2001) Measuring Market Risk with Value at Risk, John Wiley & Sons, New York.
169. Pfeifer P and Carraway R (2000) Modeling Customer Relationships as Markov Chain, Journal of Interactive Marketing, 14:43–55.
170. Pliska S (2003) Introduction to Mathematical Finance: Discrete Time Models, Blackwell Publishers, Oxford.
171. Priestley M (1981) Spectral Anslysis and Time Series, Academic Press, New York.
172. Puterman M (1994) Markov Decision Processes: Discrete Stochastic Dynamic Programming John Wiley and Sons, New York.
173. Rabiner L (1989) A Tutorial on Hidden Markov Models and Selected Applications in Speech Recognition, Proceedings of the IEEE, 77:257–286.
174. Raftery A (1985) A Model for High-order Markov Chains, Journal of Royal Statistical Society, Series B, 47:528–539.
175. Raftery A and Tavare S (1994) Estimation and Modelling Repeated Patterns in High Order Markov Chains with the Mixture Transition Distribution Model, Journal of Applied Statistics, 43: 179–199.
176. Raymond J, Michael J, Elizabeth A, Lars S (1998), A Genome-Wide Transcriptional Analysis of the Mitotic Cell Cycle. Molecular Cell, 2:65–73.
177. Richter K (1994) An EOQ Repair and Waste Disposal, In Proceedings of the Eighth International Working Seminar on Production Economics, 83–91, Igls/Innsbruch, Austria.
178. Robert C (2001) The Bayesian Choice, Springer-Verlag, New York.
179. Robinson L (1990) Optimal and Approximate Policies in Multi-period, Multi-location Inventory Models with Transshipments, Operations Research, 38:278–295.
180. Ross S (2000) Introduction to Probability Models, 7th Edition, Academic Press.
181. Saad Y (2003) Iterative Methods for Sparse Linear Systems Society for Industrial and Applied Mathematics, 2nd Edition, Philadelphia, PA.
182. Saunders A and Allen L (2002) Credit Risk Measurement: New Approaches to Value at Risk and Other Paradigms, John Wiley and Sons, New York.
183. Shahabi C, Faisal A, Kashani F and Faruque J (2000) INSITE: a Tool for Real Time Knowledge Discovery from Users Web Navigation, Proceedings of VLDB2000, Cairo, Egypt.
184. Shmulevich I, Dougherty E, Kim S and Zhang W (2002) Probabilistic Boolean Networks: a Rule-based Uncertainty Model for Gene Regulatory Networks, Bioinformatics, 18:261–274.
185. Shmulevich I, Dougherty E, Kim S and Zhang W (2002) Control of Stationary Behavior in Probabilistic Boolean Networks by Means of Structural Intervention, Journal of Biological Systems, 10:431–445.
186. Shmulevich I, Dougherty E, Kim S and Zhang W (2002) From Boolean to Probabilistic Boolean Networks as Models of Genetic Regulatory Networks, Proceedings of the IEEE, 90:1778–1792.

187. Shmulevich I, Dougherty E and Zhang W (2002) Gene Perturbation and Intervention in Probabilistic Boolean Networks, Bioinformatics, 18:1319–1331.
188. Siu T, Ching W, Fung E and Ng M (2005) On a Multivariate Markov Chain Model for Credit Risk Measurement, Quantitative Finance, to appear.
189. Siu T, Ching W, Fung E and Ng M (2005), Extracting Information from Spot Interest Rates and Credit Ratings using Double Higher-Order Hidden Markov Models, Working paper.
190. Siu T and Yang H (1999) Subjective Risk Measures: Bayesian Predictive Scenarios Analysis, Insurance: Mathematics and Economics, 25:157–169.
191. Siu T, Tong H and Yang H (2001) Bayesian Risk Measures for Derivatives via Random Esscher Transform, North American Actuarial Journal, 5:78–91.
192. Smolen P, Baxter D and Byrne J (2000) Mathematical Modeling of Gene Network, Neuron, 26:567–580.
193. Sonneveld P (1989) A Fast Lanczos-type Solver for Non-symmetric Linear Systems, SIAM Journal on Scientific Computing, 10:36–52.
194. Steward W (1994) Introduction to the Numerical Solution of Markov Chain, Princeton University Press, Princeton, New Jersey.
195. Tai A, Ching W and Cheung W (2005) On Computing Prestige in a Network with Negative Relations, International Journal of Applied Mathematical Sciences, 2:56–64.
196. Teunter R and van der Laan E (2002) On the Non-optimality of the Average Cost Approach for Inventory Models with Remanufacturing, International Journal of Production Economics, 79:67–73.
197. Thierry M, Salomon M, van Nunen J, and van Wassenhove L (1995) Strategic Issues in Product Recovery Management, California Management Review, 37:114–135.
198. Thomas L, Allen D and Morkel-Kingsbury N (2002) A Hidden Markov Chain Model for the Term Structure of Credit Risk Spreads, International Review of Financial Analysis, 11:311–329.
199. Trench W (1964) An Algorithm for the Inversion of Finite Toeplitz Matrices, SIAM Journal of Applied Mathematics 12:515–522.
200. van der Laan E (2003) An NPV and AC analysis of a Stochastic Inventory system with Joint Manufacturing and Remanufacturing, International Journal of Production Economics, 81-82:317–331.
201. van der Laan E, Dekker R, Salomon M and Ridder A (2001) An (s,Q) Inventory Model with Re-manufacturing and Disposal, International Journal of Production Economics, 46:339–350.
202. van der Laan E and Salomon M (1997) Production Planning and Inventory Control with Re-manufacturing and Disposal, European Journal of Operational Research, 102:264–278.
203. Varga R (1963) Matrix Iterative Analysis, Prentice-Hall, New Jersey.
204. Viterbi A (1967) Error Bounds for Convolutional Codes and an Asymptotically Optimum Decoding Algorithm, IEEE Transaction on Information Theory, 13:260–269.
205. Wang T, Cardiff R, Zukerberg L, Lees E, Amold A, and Schmidt E (1994) Mammary Hyerplasia and Carcinoma in MMTV-cyclin D1 Transgenic Mice. Nature, 369:669–671.
206. Wasserman S and Faust K (1994) Social Network Analysis: Methods and Applications, Cambridge Univeristy Press, Cambridge.

207. Waterman M (1995) Introduction to Computational Biology, Chapman & Hall, Cambridge.
208. White D (1993) Markov Decision Processes, John Wiley and Sons, Chichester.
209. Winston W (1994) Operations Research: Applications and Algorithms, Belmont Calif., Third Edition, Duxbury Press.
210. Wirch J and Hardy M (1999) A Synthesis of Risk Measures for Capital Adequacy, Insurance: Mathematics and Economics, 25:337–347.
211. Woo W and Siu T (2004) A Dynamic Binomial Expansion Technique for Credit Risk Measurement: A Bayesian Filtering Approach. Applied Mathematical Finance, 11:165–186.
212. Yang Q, Huang Z and Ng M (2003) A Data Cube Model for Prediction-based Web Prefetching, Journal of Intelligent Information Systems, 20:11–30
213. Yeung K and Ruzzo W (2001) An Empirical Study on Principal Component Analysis for Clustering Gene Expression Data, Bioinformatics, 17:763–774.
214. Young T and Calvert T (1974) Classification, Estimation and Pattern Recognition, American Elsevier Publishing Company, INC., New York.
215. Yuen W, Ching W and Ng M (2004) A Hybrid Algorithm for Queueing Systems, CALCOLO 41:139–151.
216. Yuen W, Ching W and Ng M (2005) A Hybrid Algorithm for Solving the PageRank, Current Trends in High Performance Computing and Its Applications Proceedings of the International Conference on High Performance Computing and Applications, August 8-10, 2004, Shanghai, China (Zhang W, Chen Z, Glowinski R, and Tong W (Eds.)) 257–264, Springer.
217. Yuen X and Cheung K (1998) Modeling Returns of Merchandise in an Inventory System, OR Spektrum, 20:147–154.
218. Zhang S, Ng M, Ching W and Akutsu T (2005) A Linear Control Model for Gene Intervention in a Genetic Regulatory Network, Proceedings of IEEE International Conference on Granular Computing, 25-27 July 2005, Beijing, 354–358, IEEE.
219. Zheng Y and Federgruen A (1991) A simple Proof for Optimality of (s, S) Policies in Infinite-horizen Inventory Systems, Journal of Applied Probability, 28:802–810.
220. http://www-groups.dcs.st-and.ac.uk/~history/Mathematicians/Markov.html
221. http://hkumath.hku.hk/~wkc/sim.xls
222. http://hkumath.hku.hk/~wkc/build.xls
223. http://www.search-engine-marketing-sem.com/Google/GooglePageRank.html.
224. http://hkumath.hku.hk/~wkc/clv1.zip
225. http://hkumath.hku.hk/~wkc/clv2.zip
226. http://hkumath.hku.hk/~wkc/clv3.zip
227. http://www.genetics.wisc.edu/sequencing/k12.htm.
228. http://www.google.com/technology/

References

Index